Charles Albert Crampton

Foods and food adulterants

Charles Albert Crampton

Foods and food adulterants

ISBN/EAN: 9783337201081

Printed in Europe, USA, Canada, Australia, Japan

Cover: Foto ©Andreas Hilbeck / pixelio.de

More available books at **www.hansebooks.com**

U. S. DEPARTMENT OF AGRICULTURE.
DIVISION OF CHEMISTRY.

BULLETIN   No. 13.

FOODS

AND

FOOD ADULTERANTS.

INVESTIGATIONS MADE UNDER DIRECTION OF

Dr. H. W. WILEY, Chief Chemist.

PART FIFTH:

BAKING POWDERS.
BY
C. A. CRAMPTON, Assistant Chemist.

PUBLISHED BY AUTHORITY OF THE SECRETARY OF AGRICULTURE.

WASHINGTON:
GOVERNMENT PRINTING OFFICE.
1889.
5360—pt. 5——1

PREFATORY NOTE.

WASHINGTON, D. C., *August* 17, 1889.

SIR: I submit herewith for your examination and approval Part Five of Bulletin No. 13 on the adulteration of food. The present part consists of an investigation of baking powders and a résumé of our present knowledge of the subject.

In these investigations we have used every endeavor to avoid error and bias. No particular powder has been favored at the expense of any other one. Our samples have been purchased in the open market and we have had them to represent as fairly as possible the character of the goods sold.

In such an investigation it is not possible to get results which will please every dealer and manufacturer, and we may therefore expect that many of our data will be distorted or denied by interested parties. A more serious embarrassment may also confront us, and that is the use of isolated portions of this report for advertising purposes.

The public official who lends the name and authority of his office for advertising purposes has little regard for either, and less for the proprieties of his position. He has, however, no longer control of the data of his analyses when they have once been published by the proper authority.

It would be well, in view of such facts, if the use of such matter for advertising purposes could be absolutely forbidden. In the present case I would like to emphasize the statement that any data or statements in the present Bulletin which may be paraded by advertisers in praise of their wares would show a discrimination wholly unauthorized by the spirit and scope of this work.

Respectfully,

H. W. WILEY,
Chemist.

Hon. J. M. RUSK,
Secretary of Agriculture.

LETTER OF SUBMITTAL.

SIR : I have the honor to submit herewith Part Fifth of Bulletin No. 13, containing the results of an investigation into the character and composition of baking powders, in continuation of the work on food adulteration. My absence in connection with the sugar experiments has greatly delayed the completion of the work and the preparation of this report.

My thanks are due to Mr. K. P. McElroy and Mr. T. C. Trescot for the intelligent assistance rendered by them in the performance of the analytical work.

Respectfully,

C. A. CRAMPTON,
Assistant Chemist.

Dr. H. W. WILEY,
Chief Chemist.

AERATION OF BREAD.

When bread is made by simply mixing flour with water and baking the dough, the result is a hard, tough, compact mass, "the unleavened bread" of the Scriptures. The use of yeast to "leaven" the dough is doubtless almost as old as the art of baking itself. Both kinds of bread are mentioned in Mosaic history, and its use was known in Egypt and in Greece at very early periods. Nothing has ever been found that could equal the action of yeast as a leavening agent. Carbonic-acid gas is generated by fermentation from the carbohydrates already existing in the bread, so that no foreign materials are introduced into it. The disengagement of the gas takes place slowly, so that it has its full effect in the lightening of the dough. This is an objection to its use, of course, when quick raising is desirable, and it is this slow action of yeast which has been the chief cause of the introduction of a chemical aerating agent.

The method of aeration invented by Dr. Danglish, in England, in March, 1859, approximates more closely the action of yeast than any other method in so far as it introduces no permanent foreign substance into the bread. In this method water which has been previously charged with carbonic dioxide is used in making up the dough, the operation begin performed in a closed vessel, under pressure. As soon as the dough is taken from this vessel it immediately rises, from the expansion of the gas contained in it. The method has been modified by using instead of water a weak wort, made by mashing malt and flour, and allowing fermentation to set in. This acid liquid absorbs the gas more readily, and perhaps has some slight effect on the albuminoids, the peptonization of which constitutes an advantage of yeast-raised bread over that made by this method, in which the aeration is purely a mechanical operation. Thus the bread made by this process is somewhat tasteless, the flavors produced by fermentation within the bread being wanting. On the other hand, there is no danger of the improper fermentations which sometimes occur, and the process is especially adapted to flours which would be apt to undergo such changes when fermented. Jago[1] says with reference to it:

Working with flours that are weak or damp or even bordering on the verge of unsoundness, it is still possible to produce a loaf that should be wholesome and palatable, certainly superior to many sodden and sour loaves one sees made from low-quality flours fermented in the ordinary manner. In thus stating that it is possible

---

[1] Chemistry of Wheat, Flour, and Bread, and Technology of Bread-making. London. 1886.

to treat flours of inferior quality by this aerating method, the author wishes specially to carefully avoid giving the impression that it is the habit of those companies which work Dauglish's method to make use of only the lower qualities of flour; he has never had any reason whatever for supposing such to be the case.

This method is in operation in all the larger cities of Great Britain, but I have no knowledge of its being used in this country.

CHEMICAL AERATING AGENTS.

The necessity of sometimes having bread preparations raised quickly for immediate baking led to the use of chemical agents for this purpose. In all of these the expansive gas is the same as where yeast is used, but instead of its being derived from the constituents of the flour, it is obtained by the decomposition of a carbonate which is introduced, together with an acid constituent to act upon it, directly into the flour. When water is added to make the dough the chemicals are dissolved, the reaction occurs, and the carbonic acid is set free, while the salt resulting from the combination of the acid with the alkaline base of the carbonate remains in the bread and is eaten with it. Many suppose, and this idea is fostered by baking-powder manufacturers, that nothing remains in the bread, that everything is driven off during the baking. This is entirely erroneous, of course, and the residue necessarily left in the bread by baking-chemicals constitutes an objection to their use, and its amount and character determine to a large extent the healthfulness of the combination used. The essential elements of such a combination are, first, a carbonate or bicarbonate which contains the gas combined with an alkaline base, and, second, an acid constituent capable of uniting with the base in the carbonate and thus liberating the carbonic-acid gas. For the alkaline constituent bicarbonate of soda, " baking-soda," is almost exclusively employed—bicarbonate of ammonia much less. For the acid constituent, however, there is great diversity in the agents used. When the housewife mixes sour milk with baking-soda to "raise" her griddle-cakes, she makes use of the free lactic acid of the former as the acid constituent of her chemical aerating agent. When she uses "cream of tartar" or acid tartrate of potassium with soda, she uses the free tartaric acid of the former as an acid constituent, and this is the same combination that is used in one class of the baking-powders sold in the market. In fact, the entire line of such powders now sold is practically the outcome of the old-time operation of domestic chemistry, mixing "saleratus" and "cream of tartar" to aerate rolls, muffins, pancakes, and such bread preparations, which were to be baked immediately after mixing, and could not well wait for the slow operation of yeast. They consist of an acid and an alkaline constituent in about the proper proportions for combination, and in a dry state, together with various proportions of a dry inert material, such as starch, added to prevent action between the chemicals themselves, so that the preparation may be kept indefinitely.

CONSUMPTION OF BAKING-POWDERS.

The quantity of the different chemical preparations made and con-sumed under the name of "baking-powders," "yeast-powders," etc., in the United States can not be stated with any degree of accuracy; neither the Statistical Division of this Department nor the Bureau of Statistics of the Treasury was able to give any information whatever upon this sub-ject. Mr. F. N. Barrett, editor of the "American Grocer," advised me that the New York Tartar Company would probably be best able to give something of an idea, at least, of the amount produced. A letter of inquiry sent to this firm elicited the following response:

DEAR SIR: Your note of inquiry of the 22d instant was received in due course of mail. We have delayed reply thereto because of the difficulty of securing with any de-gree of reliability the information you seek. We believe that no one can give a cor-rect estimate of the quantity of baking-powder annually consumed in the United States, but we are led to conclude from rather careful consideration that it amounts to between 50,000,000 and 75,000,000 pounds. Of this quantity probably two-thirds is made from cream of tartar, and the residue from phosphate and alum.

Very respectfully,

NEW YORK TARTAR COMPANY.

This would seem rather a high estimate, implying as it does an an-nual average consumption of a pound each by every man, woman, and child in the country. Probably few persons would suppose that it reached such a figure. Taking the price per pound at 50 cents, which is about the maximum retail price charged the consumer, together with the lower of the two figures given above, we would have $25,000,000 as the amount annually paid by consumers for this one article.

Granting that the above is somewhat of an overestimate, there can be little doubt that no other article which enters into the composition of food-stuffs, and which is not of itself a nutrient, is the subject of so great an expenditure.

The consumption of baking-powders does not seem to have become so extensive in Europe as in the United States, judging from the very small amount of attention bestowed upon the subject in works on food. Jago[1] makes but slight mention of their use. Doubtless the American people eat more largely of preparations of breadstuffs which are baked quickly, such as rolls, buns, etc.

In view of the large quantity of these preparations now consumed, and a lack of knowledge amongst most people concerning their compo-sition and the chemical reactions that occur in their use, I have thought it proper to give a somewhat detailed exposition of the principles in-volved, and to endeavor to explain, even to the non-scientific reader, how these powders are made, and how they act.

---

[1] Op. cit.

RECENT INVESTIGATIONS.

Two important studies of the composition and character of baking-powders have been made recently, one under the direction of the Ohio Dairy and Food Commission, and the other by the Dairy Commissioner of the State of New Jersey.[1] Work done in this way, which has the authority and weight of official sanction, is most valuable, and I have drawn largely upon the reports above mentioned in the following pages. Many other analyses of baking-powders have been made from time to time, and several extensive investigations have been carried out upon the relative merits of different kinds of powders. In fact "baking-powder literature" is quite extensive. The active competition between makers of different brands, and the methods used by them in advertising their goods, have made readers of newspapers and magazines familiar with all sorts of parti-colored statements about baking-powders in general, and certain classes and brands in particular, and unfortunately such matter is not always confined to advertising columns. Most persons know comparatively little about baking-powders, and the general ignorance on the subject is taken advantage of and intensified by the manufacturers. The analyses and testimonials of eminent chemists frequently appear in such advertisements, and are often couched in terms that do little credit to the profession. I can make no use of such publications; the only material I can accept as trustworthy are the reports cited above, where the official character of the work done affords ample assurance that the investigators were influenced by unbiased and disinterested motives. It is the proper province of such bodies as State boards of health to make investigations of this kind, and results arrived at in this way are always entitled to credence, while the conclusions of scientific men, however expert they may be, are always open to doubt when they receive compensation from parties who are interested in having the results lean in their direction.

ADULTERATION.

There is no recognized standard for the composition of a baking-powder, either in this country or abroad. To prove from a legal point of view that a powder was adulterated, it would be necessary to show that it contained some substance injurious to health. Most of the treatises on food adulteration give but little attention to this class of

---

[1] While the present publication was passing through the press, I have received another official publication upon this subject, constituting Bulletin No. 10 of the Laboratory of the Inland Revenue Department, Ottawa, Canada, and prepared by A. McGill, assistant to chief analyst. I regret that it appeared too late to allow of the incorporation into the present publication of any of the results and conclusions contained in it. Most of the powders examined were of Canadian manufacture, but the leading American brands were also included, and the analyses were quite complete.

substances, which, though not of themselves articles of food, enter into the composition of food preparations. Considerable space is devoted in such works, however, to the adulteration of bakers' chemicals. If a substance is sold as cream of tartar, for instance, which either is not cream of tartar, or is sophisticated with some cheaper substance, the seller could be convicted under food-adulteration laws, but if such a fraudulent cream of tartar were incorporated into a mixture with other chemicals and the whole sold as baking-powder, no conviction could be secured. In the famous "Norfolk baking-powder case" in England, which will be alluded to further on, the powder in question contained alum, which substance bakers are not allowed by law to use in bread. Yet the prosecution was not successful because it was directed against the sale of the powder, not against the bread made from it, there being no legal standard for substances sold as baking-powder in England.

CLASSIFICATION OF BAKING-POWDERS.

Baking-powders may be conveniently classified according to the nature of the acid constituent they contain. Three principal kinds may be recognized as follows :

(1) Tartrate powders, in which the acid constituent is tartaric acid in some form.

(2) Phosphate powders, in which the acid constituent is phosphoric acid.

(3) Alum powders, in which the acid constituent is furnished by the sulphuric acid contained in some form of alum salt.

All powders sold at present will come under some one of these heads, although there are many powders which are mixtures of at least two different classes.

TARTRATE POWDERS.

The form in which tartaric acid is usually furnished in this class is bitartrate of potassium, or " cream of tartar." Sometimes free tartaric acid is used, but not often. Bitartrate, or acid tartrate of potassium is made from crude argol obtained from grape juice. It contains one atom of replaceable hydrogen, which gives it the acidity that acts upon the carbonate. The reaction takes place according to the following equation :

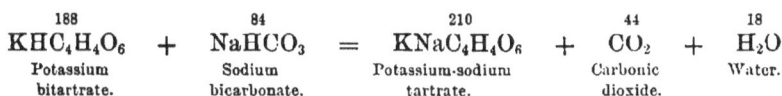

$$\underset{\substack{188 \\ \text{Potassium} \\ \text{bitartrate.}}}{KHC_4H_4O_6} + \underset{\substack{84 \\ \text{Sodium} \\ \text{bicarbonate.}}}{NaHCO_3} = \underset{\substack{210 \\ \text{Potassium-sodium} \\ \text{tartrate.}}}{KNaC_4H_4O_6} + \underset{\substack{44 \\ \text{Carbonic} \\ \text{dioxide.}}}{CO_2} + \underset{\substack{18 \\ \text{Water.}}}{H_2O}$$

It will be seen that the products of the reaction are carbonic acid and double tartrate of potassium and sodium, the latter constituting the residue which remains in the bread. This salt is generally known as Rochelle salt, and is one of the component parts of seidlitz powders.

A seidlitz powder contains 120 grains of this salt, but the crystallized salt contains four molecules of water, and thus the actual amount of crystallized Rochelle salt formed in the baking-powder reaction is greater than the combined weight of the two salts used. That is to say, if 184 grains of bitartrate and 84 grains of bicarbonate are used in a baking there will be a residue in the dough equal to 282 grains of Rochelle salt. The directions that accompany these powders generally give two teaspoonfulls as the proper amount to use to the quart of flour; probably more is generally used. This would be at least 200 grains; deducting 20 per cent. for the starch filling we have 160 grains of the mixed bitartrate and bicarbonate, and this would form 165 grains of crystallized Rochelle salt in the loaf of bread made from the quart of flour, or 45 grains more than is contained in a seidlitz powder. The popular idea is that the chemicals used in a baking-powder mostly disappear in baking, and that the residue left is very slight. I doubt if many persons understand that when they use tartrate powders, which are considered to be the best class, or at least one of the best classes of such powders, they introduce into the breadstuff very nearly an equal weight of the active ingredient of seidlitz powders, and in a loaf of bread made from it they consume more than the equivalent of one such powder.

Yet the character of this residue is probably the least objectionable of any of those left by baking-powder. Rochelle salt is one of the mildest of the alkaline salts. The dose as a purgative is from $\frac{1}{2}$ to 1 ounce. "Given in small and repeated doses it does not purge, but is absorbed and renders the urine alkaline." (United States Dispensatory.)

Free tartaric acid, used instead of the bitartrate of potassium, would give less residue. In this case the reaction would be as follows:

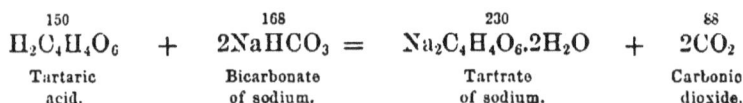

$$\underset{\substack{\text{Tartaric} \\ \text{acid.}}}{\overset{150}{H_2C_4H_4O_6}} + \underset{\substack{\text{Bicarbonate} \\ \text{of sodium.}}}{\overset{168}{2NaHCO_3}} = \underset{\substack{\text{Tartrate} \\ \text{of sodium.}}}{\overset{230}{Na_2C_4H_4O_6.2H_2O}} + \underset{\substack{\text{Carbonic} \\ \text{dioxide.}}}{\overset{88}{2CO_2}}$$

Here 150 grains of tartaric acid, with 168 grains of bicarbonate of sodium, give 230 grains of residue, or 88 grains less than the combined weight of the two ingredients. As to the character of this residue little is said in regard to the physiological properties of tartrate of sodium in the books, but probably it is essentially similar to the double tartrate. The United States Dispensatory says of it (p. 1762):

This salt, in crystals, has been recommended by M. Delioux as an agreeable purgative, almost without taste, and acting with power equal to that of the sulphate of magnesium in the dose of 10 drachms [600 grains].

I do not know why this combination should be used so seldom by baking-powder manufacturers. The free tartaric acid is more expensive than the bitartrate, but less of it is required in proportion to the amount of bicarbonate used. The former is more soluble, and this would probably be a practical objection to its use, as it is an object in baking-powders that the gas should be liberated slowly. It would perhaps be more difficult also to prevent action of the free acid upon the alkali, so that the powder would be more likely to deteriorate in keeping. Only one sample among those I examined was found to have been made with the free acid.

One obstacle formerly encountered in the manufacture of bitartrate powders was the difficulty of obtaining the bitartrate pure. It contained from 5 to 15 per cent. of tartrate of lime incident to the method of manufacture. This brought a large quantity of inert material into the powder and lowered its efficiency. Bitartrate can now be had 98 per cent. pure, quoted and guaranteed as such in the markets, so that there is no excuse for manufacturers to use the impure salt, which can properly be considered adulterated.

PHOSPHATE POWDERS.

The salt commonly used to furnish the phosphoric acid in this class is acid phosphate of lime, sometimes called superphosphate. The pure salt is monocalcic phosphate, $CaH_4(PO_4)_2$. It is made by the action of sulphuric acid upon ground bone, the result being an impure monocalcic phosphate with calcium sulphate. This mixture is sold as a fertilizer, as superphosphate. The salt is, of course, more or less purified for use in baking-powders, but the sulphate of lime is very difficult to get rid of entirely, and most phosphate powders contain considerable amounts of this impurity. The reaction which occurs when a phosphate powder is dissolved, that is the action of bicarbonate of soda upon monocalcic phosphate, is not well established, and perhaps varies somewhat with conditions. The following equation probably represents it fairly well:

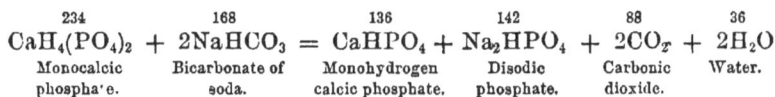

$$\underset{\substack{\text{Monocalcic} \\ \text{phospha'e.}}}{\overset{234}{CaH_4(PO_4)_2}} + \underset{\substack{\text{Bicarbonate of} \\ \text{soda.}}}{\overset{168}{2NaHCO_3}} = \underset{\substack{\text{Monohydrogen} \\ \text{calcic phosphate.}}}{\overset{136}{CaHPO_4}} + \underset{\substack{\text{Disodic} \\ \text{phosphate.}}}{\overset{142}{Na_2HPO_4}} + \underset{\substack{\text{Carbonic} \\ \text{dioxide.}}}{\overset{88}{2CO_2}} + \underset{\substack{\text{Water.}}}{\overset{36}{2H_2O}}$$

Two hundred and thirty-four grains of monocalcic phosphate combined with 168 grains of bicarbonate of soda give 136 grains of monohydrogen calcic phosphate, and 142 grams of disodic phosphate. But crystallized sodic phosphate contains twelve molecules of water, and has a molecular weight of 358. So the total amount of residue from 402 grains of the powder would be 494 grains, of which 136 grains is phosphate of lime, and the rest phosphate of soda. So we see that here also the quantity of chemicals introduced into the dough is fully equal to the amount of the baking-powder used, including filling. As to the nature of this residue in phosphate powders, it would seem to be about as unobjectionable as in the tartrates. Phosphate of soda is "mildly purgative in doses of from 1 to 2 ounces" (480–960 grains), according to the United States Dispensatory. Phosphates of calcium have the general physiological effect which is ascribed to all forms of phosphoric acid, but which does not seem to be well understood.

Phosphates are administered therapeutically in some cases of defec-

tive nutrition, and especially in scrofula, rickets, phthisis, etc. On account of their being an essential constituent of animal tissues there
would seem to be some ground for a preference over other forms of
powders. The makers of phosphate powders claim that the use of such
powders restores the phosphoric acid present in the whole grain of
wheat, which is largely removed in the bran by milling processes. This
claim would have more weight if there were not ample sources of phosphoric acid in other forms of food, and if the quantity introduced by a
baking-powder were not much greater than is required to make up the
loss in the bran, and greater than is required by the system, unless in
those cases where its therapeutic use is indicated, as in some of the conditions of malnutrition given above.

Acid phosphate of soda is said to have been used in former years as
a constituent of baking-powders, but appears to have been entirely
superseded by the lime salt.

ALUM POWDERS.

In this class the carbonic acid is set free from the bicarbonate by the
substitution of sulphuric acid, which combines with the sodium. The
sulphuric acid is furnished by some one of the general class of salts
known as alums, which are composed of a double sulphate of aluminium and an alkali metal. The alum is precipitated as hydrate, while
that portion of the sulphuric acid which was combined with it goes to
displace the carbonic acid in the bicarbonate. The alkali sulphate of
the double salt remains unchanged.

The alum of commerce is either *potash alum*, $K_2Al_2(SO_4)_4 \cdot 24H_2O$, or
ammonia alum, $(NH_4)_2Al_2(SO_4)_4 \cdot 24H_2O$, the one or the other predomi-
nating according to the relative cheapness of the alkali salt it contains. At the present time nothing but ammonia alum is met with, but
at previous periods potash alum was the salt sold exclusively as "alum."
Both salts are alike in general appearance and can not be distinguished
apart by cursory examination.

Potash alum may be made directly from some minerals, such as the
"alum stone" mined in Italy, which contain all the constituents combined. Ammonia alum, however, as well as most potash alum, is made
by the combination of the constituents obtained from different sources.
The sulphate of alumina is obtained by the action of sulphuric acid
upon pure clays, and the sulphate of ammonia from the residue of gas-
works. Solutions of the two salts in proper proportions are mixed
and the double salt obtained by evaporation and crystallization.

Crystallized potash or ammonia alum contains twenty-four molecules
of water, nearly one-half of its weight. Part of this water is lost at as
low a heat as 60° C., and it is driven off entirely, though slowly, at 100°
C. "Burnt alum" is simply alum deprived of its water of crystallization, which is generally driven off at about 200° C. Ammonia alum
decomposes at 205° C.; potash alum at a somewhat higher temperature.

Burnt alum is somewhat hygroscopic, but dissolves more slowly in water than the crystallized salt.

I have been unable to ascertain in what condition the alum is used for compounding baking-powders. Burnt alum would seem to be the form best adapted for this purpose on account of its slow solubility. Professor Cornwall says this is the form [1] used, but does not state how he obtained the information; and he states further that "crystallized alums may be used in connection with burnt alum to secure at first a more rapid escape of carbonic-acid gas." It is probable that the amount of drying given the alum used differs with different manufacturers, but it is not likely that the water of crystallization is entirely driven off.

The following equation shows the reaction taking place in a baking-powder made with burnt ammonia alum :

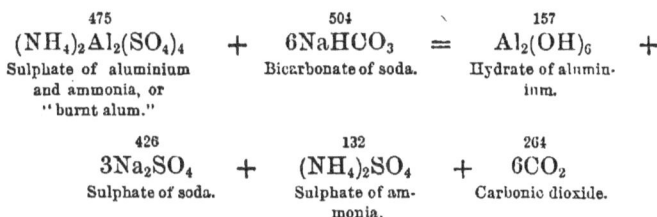

$$\underset{\substack{\text{Sulphate of aluminium}\\\text{and ammonia, or}\\\text{"burnt alum."}}}{\overset{475}{(NH_4)_2Al_2(SO_4)_4}} + \underset{\substack{\text{Bicarbonate of soda.}}}{\overset{504}{6NaHCO_3}} = \underset{\substack{\text{Hydrate of alumin-}\\\text{ium.}}}{\overset{157}{Al_2(OH)_6}} +$$

$$\underset{\substack{\text{Sulphate of soda.}}}{\overset{426}{3Na_2SO_4}} + \underset{\substack{\text{Sulphate of am-}\\\text{monia.}}}{\overset{132}{(NH_4)_2SO_4}} + \underset{\substack{\text{Carbonic dioxide.}}}{\overset{264}{6CO_2}}$$

If potash alum were used the reaction would be precisely the same with the substitution of potassium for ammonia wherever it occurs in the equation, sulphate of potash being formed instead of sulphate of ammonia.

A study of the equation will show that 475 grains of burnt alum with 504 grains of bicarbonate will produce 264 grains of carbonic acid and leave a residue consisting of 426 grains of sulphate of soda, 132 grains of sulphate of ammonia, and 157 grains of hydrate of aluminium, the last named being a precipitate insoluble in water. Sulphate of soda crystallizes with ten molecules of water, so that the total weight of residue from the 979 grains of mixed chemicals would be 1,255 grains. If a hydrated alum is used in the powder, the proportion of residue to powder would of course be less, and the proportion of gas evolved would also be less. The character of the residue is seen to be more complex than is the case with any of the classes previously discussed, and deserves special attention. The sulphate of soda is similar to other alkali salts in its physiological action. Sulphate of ammonia is not used therapeutically, but probably has an action similar to that of other ammonia salts, such as the chloride. Professor Cornwall,[2] in his report, speaks as follows concerning this point :

It is possible, however, that too little attention has been paid to the presence of ammonium salts in the residues from ammonia alum powders. * * * We do know, however, that ammonia salts, in general, are much more irritating and stimulating in their action than the corresponding soda salts, or even than the potash salts. For instance, Stillé and Maisch, speaking of ammonium bromide, state that it has a

---

[1] Report of the Dairy Commissioner of New Jersey, 1888, p. 70.   [2] Op. cit., p. 77.

more acrid taste and is more irritating than potassium bromide. Its unpleasant taste and irritating qualities render it less convenient for administration than the bromide of potassium.

We all know how mild a substance is chloride of sodium (common table salt); but of ammonium chloride Stillé and Maisch write: "The direct effects of doses of 5 to 20 grains of this salt, repeated at intervals of several hours, are a sense of oppression, warmth, and uneasiness in the stomach, some fullness in the head. If it is used for many days together in full doses, it disturbs the digestion, coats the tongue, and impairs the appetite." We have already seen how active a drug carbonate of ammonia is, and while, in the absence of proof, it would be rash to assert that sulphate of ammonia in five-grain doses is certainly injurious, yet there is abundant ground for further investigating its effect before asserting that it is milder in its effects than Rochelle salt. It may be that this question of the presence of ammonium salts in any considerable quantities in the residues of baking-powders deserves more attention than it has hitherto received.

It would seem from the above that there would be considerable difference between the physiological effects of potash and ammonia alums themselves. Yet the medical authorities make no such distinction. Ammonia alum is officinal in the British Pharmacopœia, and while the United States Pharmacopœia specifies potash alum, the particular form met with in trade is entirely determined by the comparative cheapness of manufacture.

The question of the relative harmfulness of these different salts in the residues of baking-powders is really one for the physiologist or hygienist to decide, not the chemist. Physiological experiments alone can decide them positively.

The consideration of the residue of hydrate of aluminium will be taken up later on.

POWDERS CONTAINING MORE THAN ONE ACID INGREDIENT.

As might be expected, some powders are met with which have been made up with various proportions of different acid ingredients, and which belong therefore to more than one of the above-mentioned classes. Professor Cornwall speaks as follows concerning some of these mixed powders:

The makers of alum baking-powders sometimes add tartaric acid or bitartrate to their powders, either with or without the addition of acid phosphate of lime. This is doubtless done with the best intentions, either to secure a more rapid escape of carbonic-acid gas at the outset, or otherwise improve the powder. We have found such additions in the case of several of our samples, but the presence of tartaric acid or tartrates in alum powders is very objectionable. If added in sufficient quantity to otherwise pure alum powders, they prevent the precipitation of the insoluble hydrate of aluminium entirely when the powder is boiled with water, and they may render much of the alumina soluble in water even after the bread is baked. Without doubt it would then be readily soluble in the digestive organs, producing there the effects due to alum or any other soluble aluminium compound. With one of our samples we found that the simple water solution seemed to contain as much alumina as a nitric-acid solution. In neither of these solutions could any of the alumina be thrown down by a slight excess of ammonia water, although it was readily precipitated from the solution first rendered alkaline with caustic soda, then slightly acidified with acetic acid and boiled with excess of phosphate of soda.

A case in which the character of the powder appears to be improved by such mixing, however, is furnished by the

ALUM AND PHOSPHATE POWDERS.

This combination seems to be a favorite one with manufacturers. In fact there are now comparatively few " straight " alum powders in the market, most of the cheaper grades being made of mixtures in various proportions of the alum with acid phosphate of lime. The reaction it is intended to obtain is probably the following:

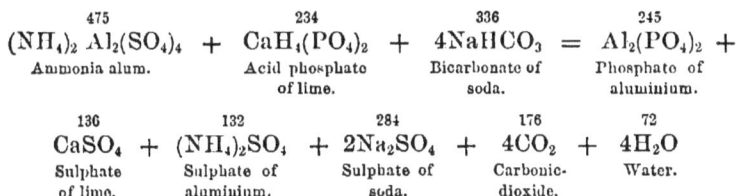

$$\underset{\substack{475 \\ \text{Ammonia alum.}}}{(NH_4)_2 Al_2(SO_4)_4} + \underset{\substack{234 \\ \text{Acid phosphate} \\ \text{of lime.}}}{CaH_4(PO_4)_2} + \underset{\substack{336 \\ \text{Bicarbonate of} \\ \text{soda.}}}{4NaHCO_3} = \underset{\substack{245 \\ \text{Phosphate of} \\ \text{aluminium.}}}{Al_2(PO_4)_2} +$$

$$\underset{\substack{136 \\ \text{Sulphate} \\ \text{of lime.}}}{CaSO_4} + \underset{\substack{132 \\ \text{Sulphate of} \\ \text{aluminium.}}}{(NH_4)_2SO_4} + \underset{\substack{284 \\ \text{Sulphate of} \\ \text{soda.}}}{2Na_2SO_4} + \underset{\substack{176 \\ \text{Carbonic-} \\ \text{dioxide.}}}{4CO_2} + \underset{\substack{72 \\ \text{Water.}}}{4H_2O}$$

If this equation be compared with the one representing the reaction in a powder made with alum alone, on page 569, it will be seen that in the former the alum goes into the residue as phosphate instead of hydrate, and the insoluble sulphate of lime takes the place of one molecule of sulphate of soda. Otherwise the reactions are similar. This reaction will only take place, of course, when the different ingredients are mixed in just the proper proportions to produce it. A number of variations may be produced by changing the relative proportions of the different ingredients.

THE "ALUM QUESTION."

The literature upon the subject of the use of alum in baking-powders, and upon the question as to its injurious effect upon the health of those who consume the bread made from it, is already quite extensive, and if quoted entire would fill a fair-sized volume. For the benefit of those who may desire to make an exhaustive study of it, I will make reference to all of the articles bearing upon the subject that have come under my observation as follows:

Alum in baking-powder, by Prof. E. G. Patrick.—*Scientific American Supplement No.* 185, 7, p. 2940.

Report of proceedings in the Norfolk baking-powder case (first trial).—*Analyst,* 4, p. 231.

Norfolk baking-powder case (second trial).—*Ibid.,* 5, p. 21.

Editorial comment on the case.—*Ibid.,* 5. pp. 13 and 34.

On the action of alum in bread making, by J. West Knights.—*Ibid.,* 5, p. 67.

Cereals and the products and accessories of flour and bread foods, by E. G. Love, Ph. D.—*Second Annual Report State Board of Health of New York,* 1882, p. 567.

On the solubility of alumina residues from baking-powders, by Lucius Pitkin.—*Journal American Chemical Society,* 9, p. 27.

Experiments upon alum baking-powders and the effects upon digestion of the residues left therefrom in bread, by Prof. J. W. Mallet.—*Chemical News,* 58, pp. 276 and 284.

As I have previously indicated, the matter of the physiological effect of the residues left by baking powders is not properly a chemical problem. On account of the interest and importance attached to it, however, it would seem necessary to give here somewhat of a résumé of the subject without attempting to arrive at a definite conclusion, or to settle, arbitrarily, the question as to whether the sale of certain forms of powders should be prohibited.

For a proper understanding of the alum question it is necessary to explain that the use of alum in bread-making is prohibited in countries having food adulteration laws, such as England and France. This is partly on account of its injurious effect upon the system, but principally because of its peculiar action, not yet well understood, in improving the color and appearance of the bread to which it has been added, so that a flour of inferior grade, or even partially spoiled, may be used to make bread which will look as well, to all appearances, as bread made from much better grades.

Blyth [1] speaks as follows of this use of alum in bread:

Alum is added to bad or slightly damaged flour by both the miller and the baker. Its action, according to Liebig, is to render insoluble gluten which has been made soluble by acetic or lactic acids developed in damp flour, and it hence stops the undue conversion of starch into dextrine or sugar. The influence of alum on health, in the small quantities in which it is usually added to bread, is very problematical, and rests upon theory more than observation. But notwithstanding the obscurity as to its action on the economy there can be no difference of opinion that it is a serious adulteration, and not to be permitted.

Allen [2] says:

Alum, or an equivalent preparation containing aluminium, is by far the most common mineral adulterant of bread, though its use has greatly decreased of late years. Its action in increasing the whiteness and apparent quality of inferior flour is unquestionable, though the cause of its influence has not been clearly ascertained. Whether there be sufficient foundation for the statements made respecting the injurious effects of alumed bread on the system is still an open question.

The following is from Hassall: [3]

With reference to the use of alum, Dr. Dauglish has written: "Its effect on the system is that of a topical astringent on the surface of the alimentary canal, producing constipation and deranging the process of absorption. But its action in neutralizing the efficacy of the digestive solvents is by far the most important and unquestionable. The very purpose for which it is used by the baker is the prevention of those early stages of solution which spoil the color and lightness of the bread whilst it is being prepared, and which it does most effectually; but it does more than needed, for, whilst it prevents solution at a time that is not desirable, it also continues its effects when taken into the stomach, and the consequence is that a large portion of the gluten and other valuable constituents of the flour are never properly dissolved, but pass through the alimentary canal without affording any nourishment whatever."

The manufacturers of alum baking-powders, however, claim that the hydrate of aluminium which is left in the residue is insoluble in the

[1] Foods, Composition and Analysis, p. 168.
[2] Commercial Organic Analysis, I, p. 371.
[3] Food, its Adulterations, and the Methods for their Detection, p. 344.

digestive juices, and therefore does not produce the effect which is attributed to the soluble forms of alum. Aluminium hydrate is insoluble in water, but readily soluble in dilute acids, especially when freshly precipitated. When heated it gradually loses its water of hydration, but does not part with it entirely short of a very high heat. When completely dehydrated it is insoluble even in dilute acid. It never reaches this condition in baked bread, in which the temperature probably never, in the center of the loaf, at least, exceeds 100° C.

Phosphate of aluminium is somewhat less soluble in dilute acids than the hydrate. In the Norfolk case an effort was made by the prosecution to show that the soluble phosphates contained in the ash of flour combined with the alum to form phosphate of aluminium, thus rendering them insoluble in the digestive juices, and depriving the flour of an important constituent, and considerable evidence was offered by the defense to show that this was not the case. Whether the addition to alum powders of sufficient acid phosphate to combine with the aluminium present as phosphate was the result of this discussion or not I can not say, but it is certain that most of the alum powders now met with are made in this way, so that if such a prosecution were to occur to day the relative position of the parties would be reversed. It would be to the interest of the alum-powder makers to show that phosphate of aluminium is insoluble in the alimentary canal. The solubility of these compounds in water or dilute acids is, of course, a question readily answered by any chemist, but their solubility in the complex and various alimentary fluids, and under the conditions of natural digestion in the human body, is quite another matter. As might be expected, the testimony which has been published upon this point is of the most conflicting character. Professor Patrick, experimenting upon cats, found little or no solution of hydrate of aluminium. Professor Pitkin, experimenting with gastric juice obtained from a dog, found some solution, although he used phosphoric acid in his powder. Professor Mallet, using an artificial gastric juice, found some solution to occur, even with the phosphate, and considerably more with the hydrate. It is not difficult to find reasons for such disagreement in results, for, besides the various character of the solvents used and the different conditions prevailing, it is easy to see that even if the hydrate and phosphate of aluminium were themselves entirely insoluble, more or less aluminium would escape the reaction, either from imperfect mixing of the powder in the dough or from improper proportioning of the different ingredients in the powder itself, so that it would go into the residue in the form of the original salt. With powders specially prepared, on the other hand, and very carefully mixed, and kneaded up thoroughly with the dough, it might be possible to find but a very little dissolved in the digestive fluids under certain conditions, even though the salts formed were slightly soluble in such fluids.

From the various evidence that has been produced on both sides of the question, I think the following conclusions may be safely drawn:

(1) That form of alum powder in which sufficient phosphate is added to combine with all the aluminium present is a better form, and less apt to bring alum into the system than where alum alone is used.

(2) It must be expected that small quantities, at least, of alum will be absorbed by the digestive fluids where any form of powder containing it is used.

(3) Whether the absorption of small quantities of alum into the human system would be productive of serious effects is still an open question, and one that careful physiological experiment alone can decide.

As the experiments made by Professor Mallet are the most recent on this subject, I quote here his conclusions. I may say that most of those based upon purely chemical work I can indorse, having confirmed many in my own work, but I think the evidence furnished by his physiological work is hardly sufficient to justify his conclusion as to the harmfulness of such powders.

GENERAL SUMMARY OF THE CONCLUSIONS REACHED.[1]

The main points which seem to be established by the experiments under discussion are, briefly stated, the following:

(a) The greater part of the alum baking-powders in the American market are made with alum, the acid phosphate of calcium, bicarbonate of sodium, and starch.

(b) These powders, as found in retail trade, give off very different proportions of carbonic-acid gas, and therefore require to be used in different proportion with the same quantity of flour, some of the inferior powders in largely increased amount to produce the requisite porosity in bread.

(c) In these powders there is generally present an excess of the alkaline ingredient, but this excess varies in amount, and there is sometimes found on the contrary an excess of acid material.

(d) On moistening with water these powders, even when containing an excess of alkaline material, yield small quantities of aluminium and calcium in a soluble condition.

(e) As a consequence of the common employment of calcium-acid phosphate along with alum in the manufacture of baking-powders, these, after use in bread-making, leave, at any rate, most of their aluminium in the form of phosphate. When alum alone is used the phosphate is replaced by hydroxide.

(f) The temperature to which the interior of bread is exposed in baking does not exceed 212° F.

(g) At the temperature of 212° F. neither the "water of combination" of aluminium hydroxide nor the whole of the associated water of either this or the phosphate is removed in baking bread containing these substances as residues from baking-powder.

(h) In doses not very greatly exceeding such quantities as may be derived from bread as commonly used, aluminium hydroxide and phosphate produce, or produced in experiments upon myself, an inhibitory effect upon gastric digestion.

(i) This effect is probably a consequence of the fact that a part of the aluminium unites with the acid of the gastric juice and is taken up into solution, while at the same time the remainder of the aluminium hydroxide or phosphate throws down in insoluble form the organic substance constituting the peptic ferment.

---

[1] Chemical News, 58, 276; also published in pamphlet form.

(*k*) Partial precipitation in insoluble form of some of the organic matter of food may probably also be brought about by the presence of the aluminium compounds in question.

(*l*) From the general nature of the results obtained, the conclusion may fairly be deduced that, not only alum itself, but the residues which its use in baking-powder leaves in bread, can not be viewed as harmless, but must be ranked as objectionable, and should be avoided when the object aimed at is the production of wholesome bread.

COMPARISON OF THE DIFFERENT CLASSES OF POWDERS IN RESPECT TO THEIR RELATIVE AERATING STRENGTH AND THE AMOUNT OF RESIDUE LEFT BY EACH.

The following comparison of the different powders described may prove interesting. It is assumed, of course, that the ingredients are combined in exactly the proper proportions, and that all the chemicals used are of full purity and strength:

Powders.	Carbonic-acid gas.	Total residue of the weight of chemicals used.
	Per cent.	*Per cent.*
Tartrate	16	104
Phosphate	22	123
Alum	27	128
Alum and phosphate	17	111

From this it will be seen that a tartrate powder, theoretically, gives the lowest percentage of carbonic-acid gas in proportion to the weight of chemicals used in its composition, together with the least weight of residue; and a straight alum powder gives the highest proportion of gas, with the greatest weight of residue. It is assumed that burnt alum is used in both the alum and the alum and phosphate powders. The residues are calculated to hydrated salts in all cases. No account is made of inert "filling," as that would be the same in each case. It should of course be remembered that in the above calculation the *total weight* of residue is reckoned in each case without regard to solubility or relative effect upon the system of the various salts formed. This has been sufficiently discussed under the different classes.

CARBONATE OF AMMONIA.

This salt is used to some extent as an ingredient in baking-powders. It is also often used alone by bakers as a chemical aerating agent. It does not necessarily require an acid to set free its gases, being volatilized without decomposition simply in heating. The commercial salt, familiar to everybody as "smelling-salts," or *sal volatile*, is obtained by subliming a mixture of two parts of chalk and one part of sal ammoniac or sulphate of ammonia. The salt is then resublimed with the addition of some water, and a white semi-transparent mass is obtained, which has a strongly ammoniacal smell, and a pungent, caustic taste. It has the composition $N_3H_{11}C_2O_5$, and consists of a compound of hydrogen ammonium carbonate with ammonium carbonate, $H(NH_4)CO_3 + NH_4CO_2$ NH_2. "When heated the salt is wholly dissipated, without charring; if the aqueous solution is heated to near $47°C$. it begins to lose carbonic-acid gas, and at $88°$ it begins to give off vapor of ammonia." (United States Pharmacopœia.) The question of the propriety of the use of this salt in baking does not seem to have received a great deal of attention, and opinions differ. Hassall[1] says of it :

> * * * Of these, by far the best is carbonate of ammonia ; this is a volatile salt, and its great advantage is that it is entirely or almost entirely dissipated by the heat employed in the preparation of the bread ; and thus the necessary effect is produced without risk of injurious results ensuing.

This would doubtless hold good if it were quite certain that the salt is *entirely* driven off by the baking of the bread, for it is a very active therapeutic agent, acting as a corrosive poison when taken in sufficient amount. The ordinary dose is five grains. Doubtless in the small quantities used in baking-powders, and in the presence of other chemicals, there is little danger of its being left in the bread undecomposed, but the advisability of its use alone as an aerating agent is open to grave doubt.

COMPOSITION OF BAKING-POWDERS AS FOUND IN THE MARKET.

The following analyses and discussion, by Prof. H. A. Weber, form a part of the Annual Report of the Ohio State Dairy and Food Commissioner for 1887 :

BAKING-POWDERS.

Much complaint has been made to this commission of the character of the baking-powders of commerce. It was believed by many that there was a great deal of adulteration and impurity in the ordinary baking-powders used by our people, and that the public health was seriously affected thereby. Recognizing the importance of this matter to the health and domestic economy of the people of the State, I gave public notice of my purpose to investigate the purity and healthfulness of the various brands of the baking-powders of commerce. I sought all possible information on this subject, and collected and submitted to analysis by the State chemist thirty brands of baking-powders, such as I found on sale in many sections of the State. In September the result of these investigations and analyses was published in an official circular for the benefit of the consumers of this class of goods. There was no thought, wish, or purpose upon the part of this commission to aid or to defeat the enterprise of any manufacturer of these goods. Indeed, we had not any possible intimation as to what the analysis would show in any particular brand until the work was accomplished. There was simply the impartial purpose to inform the public as to the chemical composition of the several brands sold by the trade throughout the State, so that with the knowledge of the facts they might not claim that they were being defrauded or imposed upon, but be able intelligently to choose the goods they deemed most health-

---

ful aud desirable. This circular with its analysis has attracted so much attention throughout the State and country and is of such significance as to demand a place in this report, and it is therefore given here in full.

CIRCULAR No. 6.—BAKING-POWDERS.

This commission has been for some months investigating the baking-powders of commerce most generally used and sold in this State. And we herein submit to the people of the State the result of that investigation.

We have analyzed thirty brands of baking-powder, seeking those brands which were apparently most generally sold throughout the State. We submit the result of these analyses to the people who are the consumers of such goods that they may know the true chemical character of these several varieties.

It is generally supposed that there is a vast deal of " adulteration " in baking-powder, but since there is at law no standard of excellence or purity in baking-powder, it is difficult to say what is an adulteration, unless it be an unhealthful ingredient.

As a matter of fact, any powdered composition that is healthful and which in solution in moist dough will generate carbonic-acid gas and " raise " bread, or cause it to be porous and light, may be properly called a baking-powder. And accordingly we find very many varieties or brands of baking-powders on the market made from widely-different materials.

The best baking-powder is, of course, that in which (the ingredients being healthful) the largest amount of carbonic gas is generated to the spoonful of powder, and the least amount and least hurtful character of the resultant salt remains in the bread.

For an intelligent view of the whole field we classify these varieties into three general divisions. In each of these the active agents of the compound are kept dry, and thus free from fermentation in the package, by the use of a given per cent. of starch, wheat flour, or rice flour. These are used simply as dry filling to keep the chemical agents from acting upon each other in the package.

In this classification we have—

(1) Cream of tartar baking-powders.
(2) Phosphate baking-powders.
(3) Alum baking-powders.

The chemical formula and the percentages of the active agents vary with each brand. But generally stated we have in the

FIRST CLASS.

Cream of tartar, } Changed by chemical action in the dough to the double salt of
Bicarbonate of soda, } tartrate of potassium and sodium, and carbonic-acid gas.
Starch or flour, }

The cream of tartar and bicarbonate of soda, dissolved by the water or moisture in the dough, unite chemically and form in the bread the double salt of tartrate of potassium and sodium, and carbonic-acid gas, the latter escaping in the baking heat.

SECOND CLASS.

Acid calcium phosphate, } Changed by chemical action into calcium phosphate,
Bicarbonate of soda, } sodium phosphate, and carbonic-acid gas.
Starch or flour, }

THIRD CLASS.

Ammonium alum, } Changed by chemical action into hydrate of aluminium, so-
Bicarbonate of soda, } dium sulphate, ammonium sulphate, and carbonic-acid gas.
Starch or flour, }

In some brands of the cream of tartar baking-powder a small per cent. of carbonate of ammonia is used; but this is considered to be too small an amount to be hurtful. There is a prevalent belief created by the erroneous statement of manufacturers, that the salts from which carbonic-acid gas is generated pass off in the form of escaping gas, scarcely leaving a trace of their presence in the bread. But this is not true. These resultant salts formed by the chemical action in the dough remain in the bread, while the gas generated by such chemical action, and which is but a small per cent. of the whole, alone passes off in the process of baking.

From this fact many persons condemn the entire class of alum baking-powders as being unhealthful. Pure alum is undoubtedly a hurtful salt, and the resultant salts from its combination with soda can scarcely be less hurtful. And yet this is a question about which "doctors disagree;" any number of conflicting opinions and certificates can be had from eminent chemists on either side of this question.

The official investigation of this class of baking-powders made in England to test their healthfulness resulted in their favor. But this conclusion rested upon the statement of chemists that the resultant salt of hydrate of aluminium remaining in the bread was insoluble, and hence unhurtful when taken into the stomach. But some of the ablest chemists of this country declare that hydrate of aluminium is quite soluble, and hence is as hurtful as the alum in other forms. So that the question is yet an open one to be determined by further careful scientific investigation.

As to the general healthfulness of cream of tartar and phosphate baking-powders when properly used, there is but little difference of opinion; but an intelligent knowledge of their strength and freshness and of the manner and rapidity of the chemical combinations in the process of bread-making is necessary to the baker in order to insure good bread.

These thirty brands were analyzed very carefully by Prof. H. A. Weber, State chemist at Columbus, Ohio, and are such as are generally sold throughout the State. The condition of some of these brands was not such as to show them at their best advantage, since some were old while others were fresh. But since they were bought in the open market without discrimination, they fairly present what the consumer must buy.

S. H. HURST,
Ohio Dairy and Food Commissioner.

General S. H. HURST,
     Ohio Dairy and Food Commissioner:

SIR: The following is a complete report of the analyses of baking-powders received June 3 and July 7, 1887:

The list, as will be seen from the analyses, includes three kinds of baking-powders, in which the acid principle is respectively cream of tartar, acid phosphate of calcium, and alum.

The carbonic acid evolved with water was determined with great care, and from this the amount of bicarbonate of soda and the acid principle was calculated according to well-known reactions.

The starch, or as in some cases the flour, was determined directly, due allowance being made in case of the alum powders for the loss of water of the aluminium hydroxide in the dried residue upon ignition.

Accidental impurities of commercial drugs were not taken into account, as they would be very small and unimportant in case of the alum powder, while in the cream of tartar powder the ingredients used were found in the course of analysis to be practically pure.

The excess of bicarbonate of soda or of the acid principle was determined volumetrically and added, as the case might be, to the results obtained by calculation.

Respectfully submitted.

H. A. WEBER,
Chemist.

ANALYSES OF BAKING-POWDERS.

CREAM OF TARTAR BAKING-POWDERS.

1. Royal.

Carbonic-acid gas, 11.80 per cent.

Bicarbonate of soda	25.21
Cream of tartar	50.44
Starch	17.10
Tartrate of potassium and sodium, moisture, etc	7.25

100.00

This powder contained a small percentage of ammonium carbonate, which was calculated as bicarbonate of soda above.

2. Dr. Price's.

Carbonic-acid gas, 10.50 per cent.

Bicarbonate of soda	21.14
Cream of tartar	44.90
Starch	21.30
Tartrate of potassium and sodium, moisture, etc	12.66

100.00

3. *Pearson's.*

Carbonic-acid gas, 11.60 per cent.

Bicarbonate of soda	23.24
Cream of tartar	49.57
Starch	12.60
Tartrate of potassium and sodium, moisture, etc	14.39
	100.00

This sample contained ammonium carbonate.

4. *Cleveland's.*

Carbonic-acid gas, 12.80 per cent.

Bicarbonate of soda	26.12
Cream of tartar	54.70
Starch	9.00
Tartrate of potassium and sodium, moisture, etc	10.18
	100.00

5. *Snow Drift.*

Carbonic-acid gas, 10.60 per cent.

Bicarbonate of soda	20.24
Cream of tartar	48.62
Starch	13.60
Tartrate of potassium and sodium, moisture, etc	17.54
	100.00

6. *Upper Ten.*

Carbonic-acid gas, 11.30 per cent.

Bicarbonate of soda	21.57
Cream of tartar	48.31
Starch	20.90
Tartrate of potassium and sodium, moisture, etc	9.22
	100.00

7. *De Land's.*

Carbonic-acid gas, 10.00 per cent.

Bicarbonate of soda	19.09
Cream of tartar	48.39
Starch	0.00
Tartrate of potassium and sodium, moisture, etc	32.52
	100.00

This powder contained no filling.

8. *Sterling.*

Carbonic-acid gas, 11.00 per cent.

Bicarbonate of soda	21.84
Cream of tartar	47.03
Starch	18.50
Tartrate of potassium and sodium, moisture, etc	12.63
	100.00

PHOSPHATIC BAKING-POWDERS.

9. *Horsford's.*

Carbonic-acid gas, 13.00 per cent.

Bicarbonate of soda	27.34
Free phosphoric acid	14.47
Starch	21.80
Insoluble ash (calcium phosphate, calcium carbonate, etc.)	19.50
Sodium phosphate, moisture, etc	16.89
	100.00

10. *Wheat.*

Carbonic-acid gas, 4.00 per cent.

Bicarbonate of soda	9.32
Free phosphoric acid	4.45
Starch	0.00
Insoluble ash (calcium phosphate, calcium carbonate, etc.)	26.90
Sodium phosphate, moisture, etc.	59.33
	100.00

This sample contained no filling and was badly caked.

ALUM BAKING-POWDERS.

11. *Empire.*

Carbonic-acid gas, 5.80 per cent.

Bicarbonate of soda	11.08
Alum	10.41
Starch	44.25
Hydrate of alumina, sodium sulphate, ammonium sulphate, moisture, etc	34.26
	100.00

12. *Gold.*

Carbonic-acid gas, 6.70 per cent.

Bicarbonate of soda	13.63
Alum	12.03
Starch	44.00
Hydrate of alumina, sodium sulphate, ammonium sulphate, moisture, etc	20.34
	100.00

13. *Veteran.*

Carbonic-acid gas, 6.90 per cent.

Bicarbonate of soda	14.66
Alum	12.13
Starch	49.85
Hydrate of alumina, sodium sulphate, ammonium sulphate, moisture, etc	23.36
	100.00

14. *Cook's Favorite.*

Carbonic-acid gas, 5.80 per cent.

Bicarbonate of soda	11.92
Alum	10.41
Starch	42.75
Hydrate of alumina, sodium sulphate, ammonium sulphate, moisture, etc	34.92
	100.00

15. *Sunflower.*

Carbonic-acid gas, 6.30 per cent.

Bicarbonate of soda	14.44
Alum	11.31
Starch	38.65
Hydrate of alumina, sodium sulphate, ammonium sulphate, moisture, etc	35.60
	100.00

16. *Kenton.*

Carbonic-acid gas, 6.20 per cent.

Bicarbonate of soda	12.59
Alum	11.14
Starch	38.10
Hydrate of alumina, sodium sulphate, ammonium sulphate, moisture, etc	38.17
	100.00

17. Patapsco.

Carbonic-acid gas, 6 per cent.

Bicarbonate of soda	12.30
Alum	10.77
Starch	36.85
Hydrate of alumina, sodium sulphate, ammonium sulphate, moisture, etc	40.08
	100.00

18. Jersey.

Carbonic-acid gas, 10.40 per cent.

Bicarbonate of soda	20.70
Alum	19.05
Starch	44.20
Hydrate of alumina, sodium sulphate, ammonium sulphate, moisture, etc	16.05
	100.00

19. Buckeye.

Carbonic-acid gas, 6.90 per cent.

Bicarbonate of soda	13.82
Alum	12.13
Starch	44.20
Hydrate of alumina, sodium sulphate, ammonium sulphate, moisture, etc	29.85
	100.00

20. Peerless.

Carbonic-acid gas, 7 per cent.

Bicarbonate of soda	14.21
Alum	12.94
Starch	46.57
Hydrate of alumina, sodium sulphate, ammonium sulphate, moisture, etc	26.28
	100.00

21. Silver Star.

Carbonic-acid gas, 6.90 per cent.

Bicarbonate of soda	14.66
Alum	12.13
Starch	41.33
Hydrate of alumina, sodium sulphate, ammonium sulphate, moisture, etc	31.88
	100.00

22. Crown.

Carbonic-acid gas, 8.40 per cent.

Bicarbonate of soda	16.88
Alum	15.08
Starch	51.35
Hydrate of alumina, sodium sulphate, ammonium sulphate, moisture, etc	16.69
	100.00

23. Crown (marked "Special").

Carbonic-acid gas, 8.60 per cent.

Bicarbonate of soda	18.10
Alum	15.44
Starch	41.37
Hydrate of alumina, sodium sulphate, ammonium sulphate, moisture, etc	25.09
	100.00

24. One Spoon.

Carbonic-acid gas, 5.75 per cent.

Bicarbonate of soda	12.66
Alum	10.33
Starch	18.33
Hydrate of alumina, sodium sulphate, ammonium sulphate, moisture, etc	58.68
	100.00

25. *Wheeler's No. 15.*

Carbonic-acid gas, 11.35 per cent.

Bicarbonate of soda	22.51
Alum	20.38
Starch	29.38
Hydrate of alumina, sodium sulphate, ammonium sulphate, moisture, etc	27.73
	100.00

26. *Carlton.*

Carbonic-acid gas, 6.60 per cent.

Bicarbonate of soda	13.44
Alum	11.85
Starch	43.77
Hydrate of alumina, sodium sulphate, ammonium sulphate, moisture, etc	30.94
	100.00

27. *Gem.*

Carbonic-acid gas, 8.45 per cent.

Bicarbonate of soda	16.13
Alum	15.17
Starch	32.13
Hydrate of alumina, sodium sulphate, ammonium sulphate, moisture, etc	36.57
	100.00

28. *Scioto.*

Carbonic-acid gas, 8.80 per cent.

Bicarbonate of soda	16.80
Alum	15.80
Starch	49.15
Hydrate of alumina, sodium sulphate, ammonium sulphate, moisture, etc	18.25
	100.00

29. *Zipp's Grape Crystal.*

Carbonic-acid gas, 10.90 per cent.

Bicarbonate of soda	22.49
Alum	19.57
Starch	45.95
Hydrate of alumina, sodium sulphate, ammonium sulphate, moisture, etc	11.99
	100.00

30. *Forest City.*

Carbonic-acid gas, 7.80 per cent.

Bicarbonate of soda	15.73
Alum	13.63
Starch	46.60
Hydrate of alumnia, sodium sulphate, ammonium sulphate, moisture, etc	24.04
	100.00

Since the issuance of the foregoing circular the manufacturers of certain of these brands of baking-powders have sought to pervert the facts brought out by these analyses, and, by arguments and conclusions wholly unwarranted by the facts stated, or by the principles of chemical science, have for their own benefit held and assumed that the circular and analyses show a state of facts which they do not show, and lead to conclusions to which they do not lead. Nevertheless, the truth will assert itself, and this investigation and discussion, which is still going on, will throw a flood of light on this whole field of commercial food products that will be of incalculable benefit to the people of the State.

Professor Weber's analyses are rather superficial and incomplete, probably being made under conditions that did not admit of thorough quantitative work. He has overlooked entirely, for instance, the fact of the presence of phosphoric acid in many alum powders.

PROFESSOR CORNWALL'S REPORT ON BAKING-POWDERS.

Following are the results of an examination of a large number of baking-powders by Prof. H. B. Cornwall, together with his description of the methods of analysis employed, and his observations and conclu sions.[1]

METHODS OF ANALYSIS.

The analysis was directed toward determining the "strength" of the powders, or their yield of carbonic-acid gas, and their composition, so far as to indicate the nature of the chief active constituents. No great importance was attached to the amount of starch or other legitimate "filling," which only has an effect on the strength of the powder, nor was it possible to examine so large a number of samples minutely as to the residues left by them. Especial attention was therefore paid to the presence of possible objectionable constituents of the residues, and to ingredients that might render the use of the powders injurious.

Carbonic-acid gas.—This was determined with great care by boiling 1 gram (15.43 grains) of the powder with 125 to 130 cubic centimeters (about $4\frac{1}{4}$ fluid ounces) of distilled water in a roomy flask, connected with a Classen condensing, drying, and absorbing apparatus (Classen, Quantitative Chemische Analyse, 1885), the carbon dioxide being absorbed in soda-lime tubes, of which there were two, having their further ends charged with carefully-dried chloride of calcium. The contents of the flask were boiled, with proper use of a slow current of air, for one and one-half to one and three-quarters hours, and the current of air was kept up for half an hour after removing the flame, so that the whole operation lasted from two to two and one-half hours. Only in this way was the carbonic-acid gas with certainty to be expelled from the somewhat viscid, starchy water solution and completely carried over into the absorption tubes.

Tested before the analyses were begun, with a sample of probably one of the best commercial bicarbonates of soda in the market, the absorption apparatus yielded 51.38 per cent. of carbon dioxide; measurement of the gas (see below) indicating 51.44 per cent.

Tested in the course of the series of analyses by decomposing Iceland spar (crystallized carbonate of lime) with citric acid in the presence of starch, in the proportion used in the average of good cream of tartar baking-powders, the absorption apparatus showed 43.83 per cent., theory requiring, for absolutely pure carbonate of lime, 44 per cent.

As a check analysis, when it could be properly done, the gas evolved from the powder by 10 cubic centimeters of a mixture of one volume of hydrochloric acid, specific gravity 1.2, with two volumes of water, in a Scheibler's evolution bottle, was collected over mercury and measured, correction being made for atmospheric pressure, temperature, and moisture, and also an allowance for the carbon dioxide retained by acid of the strength used, as determined by tests with the Iceland spar. Enough baking-powder was taken to give 90 to 110 cubic centimeters of gas. The results by measurement averaged 0.12 per cent. below those by absorption and weighing of the gas, probably on account of the difficulty of liberating the gas, even by violent shaking, from the somewhat viscid liquid produced by the action of the strongly acid solution on the starch of flour. The greatest difference by the two methods was 0.29 per cent. Whenever a sample showed a rather low percentage of carbonic-acid gas, or left a

---

[1] Report of the Dairy Commissioner of New Jersey, 1888, p. 82.

decidedly alkaline solution, duplicate tests were made by the soda-lime absorption, and no dependence was placed on measurement, but in other cases it was a most convenient and reliable check on the other method.

Sulphates.—These were detected in the cold-water solution of the baking-powder, bearing in mind the possible solvent action of citric acid on the barium sulphate. No attention was paid to minute quantities of soluble sulphates.

Ammonia salts.—These were detected by rubbing the powders with water and slaked lime, after ascertaining that ordinary samples of flour gave no reaction for ammonia under the conditions of our tests. No notice was taken of ammonia unless the turmeric paper was rapidly and decidedly colored.

Phosphates.—It was found that even in the presence of tartaric acid these could generally be detected by means of ammonium molybdate in the solution of the powder in very dilute nitric acid. In cases of doubt, the powder was first fused with carbonate of soda and nitrate of potash.

Alumina.—Although it could always be detected in the solution of the powder in very dilute nitric acid, at least, by the aid of acetic acid and phosphate of soda, yet all of the powders were also tested by fusion with carbonate of soda and nitrate of potash, extraction with boiling water, acidifying the filtered solution with hydrochloric acid and precipitating with ammonia water. The alumina in the precipitate was identified as such, however obtained. During the fusion abundant evidence of the presence of iron compounds as an impurity in the alum powders was frequently obtained, showing carelessness or ignorance on the part of the makers.

Tartaric acid and tartrates.—Free tartaric acid was dissolved out by absolute alcohol and identified. Tartrates were systematically tested for in case of doubt, but, in general, it was deemed sufficient to confirm their presence by shaking the powder with ammonia water, filtering, adding a crystal of nitrate of silver, and heating gently to form the characteristic silver mirror. It was found that phosphates and citrates did not interfere with this test when any considerable quantity of tartrates was also present in the solution, but it was depended on only as confirmation of the presence of tartrates in the cream of tartar powders.

Potash.—This was detected by holding some of the powder on a platinum wire in the Bunsen-burner flame and observing the flame coloration through a solution of permanganate of potash so strong as scarcely to transmit diffused daylight. Unless a decided red flame coloration was obtained, potash was certainly absent in any notable quantity.

RESULTS OF ANALYSIS.

The following tables give the results of the analysis of our samples, so far as was necessary to classify them and determine their "strength," that is, the percentage of carbonic-acid gas. The cubic inches of gas are given from 1 ounce avoirdupois of powder, at a temperature of 60° F., and barometer at 30 inches:

I. *Cream of tartar powders.*—In this class are placed all powders giving reactions for tartaric acid and potash, and free from phosphates, alumina, and any considerable quantity of soluble sulphates. Ammonia was sometimes present; whether as sesquicarbonate or bitartrate was not determined. Free tartaric acid was found in one case. Its presence has no effect on the wholesomeness of the powder, nor has the small amount of ammonia in any case found. The writer's experience is that the powders free from ammonia compounds yield just as light biscuits, etc., as the others.

As regards purity of materials, there seems little choice between the higher grades of these powders.

II. *Acid phosphate of lime powders.*—The first two of these were packed in tightly-corked glass bottles, and contained enough starchy material to keep them from deteriorating in these bottles.

The bread preparation consisted of two separate powders, each in a paper package. One was bicarbonate of soda, the other acid phosphate of lime mixed with starch. The strength was determined on a mixture of the two in the proportions directed on the packages.

The wheat powder was put up in tin boxes, without starch or other filling. One sample was in excellent order, the other much caked.

III. *Alum and phosphate powders.*—This class embraces powders showing ammonia, soluble sulphates, alumina, and phosphates, when tested as already described.

A few showed potash reactions, and in some there was evidence of tartaric acid or some other substance reducing silver abundantly from ammoniacal solutions. In such cases, of course, potash alum and bitartrate of ammonium may have been present, or the reactions may have been caused by cream of tartar, or by free tartaric acid. The possible combinations are very numerous, and the analysis, however complete, will not always indicate the exact combination. Inasmuch as some of these powders showed considerable alumina in the simple water solution, a more detailed examination of them is recommended, for the reasons already given. The actual presence of acid phosphate of lime, or of any other acid phosphate, was not proven, but all contained some phosphate, and have therefore been classed as indicated, although probably in every case they were made with acid phosphate of lime.

As already mentioned, the low grade of several is, perhaps, from deterioration, due to the presence of the acid phosphate in packages not sufficiently air-tight. Acid phosphate will not keep well when mixed with bicarbonate of soda, except in well-corked bottles. Tin cases are not tight enough.

Many of these powders contained sulphate of lime, chemically equivalent to *terra alba*. This was, perhaps, in no case added as an adulterant, but was a part of the acid phosphate of lime used; the latter not having been separated from the sulphate of lime formed in its manufacture. The presence of this sulphate of lime must be regarded as objectionable. None of these powders are as strong as they might be made, and most of them are very deficient in strength. Apart from questions of healthfulness, there can be no economy in buying some of these powders.

IV. *Alum powders.*—Here are classed the powders showing the same reactions as the preceding class, but free from phosphates. All appeared to be ammonia alum powders, but reactions for potash and tartaric acid were not wanting among them. Only one of them begins to come up to the strength which a "straight" burnt ammonia alum powder might have.

V. *Unclassed powders.*—The composition of these will be indicated under the special remarks.

I.—*Cream of tartar powders.*

No.	Brand.	Carbonic-acid gas.	Cubic inches carbonic-acid gas per ounce.	Remarks.
		Per cent.		
4	The Best	11.60	107.3	
4	Sea Foam	10.86	100.5	Yields a little ammonia and soluble sulphate.
23	Sterling	11.70	108.2	Yields ammonia reactions.
29	Health	6.96	64.44	Final reaction of aqueous solution strongly alkaline. See special remarks.
50	Health	7.25	67.1	
39	None Such	12.64	116.9	
40	Cleveland's	13.27	122.7	Received in June.
43	Cleveland's	13.82	127.8	Received in November.
41	Royal	13.56	125.43	Yields ammonia reactions. Received in May.
42	Royal	13.06	120.8	Yields ammonia reactions. Received in November.
45	Price's "Cream"	11.95	110.5	Received in May. Contains free tartaric acid.
53	Price's "Cream"	12.20	112.9	Received in December.
	Average, 8 brands	11.60		Excluding 29 and 50, average is 12.46 per cent. of carbonic acid.

II.—*Acid phosphate of lime powders.*

No.	Brand.	Carbonic-acid gas.	Cubic inches carbonic-acid gas per ounce.	Remarks.
		Per cent.		
46	Horsford's Phosphatic	14.95	138.3	Received in August. In 8-ounce glass bottle.
54	Horsford's Phosphatic............	14.01	129.6	In retail dealer's stock one year. A little gas escaped on opening the 4-ounce bottle.
47	Rumford's Yeast Powder.........	13.51	125.0	Received in May. In 8-ounce glass bottle.
48	Rumford's Yeast Powder...	13.89	128.5	Received in August. In 8-ounce glass bottle.
49	Horsford's Bread Preparation....	15.39	142.4	Received in August. Bi-carbonate soda and acid phosphate put up in separate papers. The acid phosphate was not quite free from soluble sulphates.
21	Wheat............................	15.62	144.5	In tin box; in good order.
52	Wheat............................	5.83	53.9	In tin box; much caked.

NOTE.—Since the rapidity with which a baking-powder gives off carbonic-acid gas in the cold varies with the ingredients used, it was deemed worth while to test some powders as follows: Forty-five grains (3 grams) of each was mixed with as little shaking as possible with $\frac{1}{6}$ ounce (5 cubic centimeters) of water, and the volume of gas evolved in five minutes was measured.

	Per cent.
Cleveland's yielded of its carbonic acid..	49.6
Royal yielded of its carbonic acid...	45.6
Horsford's yielded of its carbonic acid...	68.8
A "straight" burnt alum powder yielded of its carbonic acid..................	6.3

III.—*Alum and phosphate powders.*

No.	Brand.	Carbonic-acid gas.	Cubic inches carbonic-acid gas per ounce.	Remarks. (All give ammonia reactions.)
		Per cent.		
1	Patapsco	8.32	77.0	
2	Washington	8.81	82.5	Received in May.
27	Washington	9.97	92.2	Received in November.
3	Davis's O. K	8.99	83.2	
7	McDowell's G. and J............	9.70	89.7	
9	Lincoln	9.73	90.0	
10	Kenton	7.01	64.8	Received in October. Another sample received in May gave 3.81 per cent.
11	State	6.70	62.0	Received in October.
15	State	8.42	77.9	Received in May.
13	On Top..........................	9.17	84.8	
16	Perfection	5.09	47.1	In pasteboard box with tin ends.
19	Silver Star	9.51	88.0	
24	Our Own	10.47	96.8	
35	White Star	10.09	93.3	
28	Somerville......................	8.39	77.6	
30	Grape	10.02	92.7	
31	Sovereign	8.96	82.9	
32	A. and P. (Atlantic and Pacific)...	8.97	83.0	
33	Higgins	6.63	61.3	Received in September.
51	Higgins	11.30	104.5	Received in December.
34	Windsor.........................	8.77	81.1	
37	Brooks and McGeorge...........	10.16	94.0	
38	Henkel's	10.24	94.7	
	Average, 20 brands	8.97	....	

IV.—*Alum powders.*

No.	Brand.	Carbonic-acid gas.	Cubic inches carbonic-acid gas per ounce.	Remarks. (All show ammonia reactions.)
		Per cent.		
8	Miles's "Prize"....................	9. 63	89. 1	Shows potash reactions and reduces silver abundantly.
20	Four Ace..........................	10. 31	95. 4	
26	Feather Weight	9. 63	89. 1	
36	One Spoon.........................	16. 77	155. 1	Two other samples gave respectively 15.35 and 16.73 per cent.

V.—*Unclassed powders.*

No.	Brand.	Carbonic-acid gas.	Cubic inches carbonic-acid gas per ounce.	Remarks.
		Per cent.		
6	Silver Prize	8. 14	75. 3	Shows potash and ammonia reactions, and reduces silver abundantly.
22	Orange	8. 00	74. 0	Contains a soluble alumina compound. See special remarks.
18	Our Best	6. 15	56. 9	Shows ammonia reactions; contains much soluble sulphates. and some free tartaric acid.

SPECIAL REMARKS.

Sample No. 6.—This sample shows strong reactions for ammonia, soluble sulphates, soda, and potash. Its aqueous solution, rendered ammoniacal before filtering, reduces silver from a crystal of the nitrate, as a bright coating on the glass. It would have been classed among the cream of tartar powders had it not shown altogether too much soluble sulphates. Shaken with absolute alcohol it renders this slightly alkaline; boiled with water the powder gives a decidedly alkaline solution. It contains some alumina.

Sample No. 18.—This gave strong reactions for ammonia, soluble sulphates, soda, and potash. Tartaric acid was extracted from it by shaking with absolute alcohol. It may contain some cream of tartar, but has too much soluble sulphates to warrant placing it among the cream of tartar powders.

Sample No. 22.—This powder gave strong reactions for alumina, soluble sulphates, and soda. Neither potash nor ammonia was present. The label stated that it contained grape (tartaric?) and orange (citric?) acids, combined "with natron, bicarb. soda, and corn starch," and the analysis indicated the presence of tartrates and citrates, as well as much alumina, which was abundantly found in the aqueous solution of the powder even after boiling with the undissolved residue. Alumina in a soluble form was also extracted in considerable quantity by water alone from bread made with this powder. Apparently, the organic acids kept it in soluble condition. Since neither potash nor ammonia was present, the alumina appears to have been added in the shape of sulphate of alumina, or else of soda alum.

After washing away the starchy matter with chloroform and examining the residue under the microscope in polarized light, crystalline fragments of a singly refracting substance were observed in abundance, together with doubly refracting crystalline materials. Although no octahedral crystals could be distinctly made out, yet there were some fragments of soda alum, which is as good as any other alum for making baking-powders, so far as chemical and physiological effects are concerned. It is more likely to be affected by moisture than "burnt alum."

Sample No. 29.—This powder was very strongly alkaline, containing so great an excess of bicarbonate (or carbonate) of soda that if the proper amount of cream of tartar had been used the powder would have yielded about 11 per cent. of carbonic-acid gas.

CONCLUSIONS.

Our investigations show that while especially the higher grades of cream of tartar and acid phosphate of lime powders are maintained at a quite uniform standard of excellence, the State is flooded, also, with many baking-powders of very poor quality—cheap goods, poorly made. Of the thirty-nine brands examined, twenty-five contain alum or its equivalent, in the shape of some soluble alumina compound; eight are cream of tartar powders, with small quantities of other ingredients in several cases; four are acid phosphate of lime powders; two belong properly under none of the above classes.

With one exception, the powders containing alum all fall below the average strength of the cream of tartar powders, and in the majority of cases they fall much below the better grades of the cream of tartar powders.

In the cream of tartar and the acid phosphate of lime powders, no indications of substances likely to be injurious to health, in the quantities used, have been found.

More evidence against the use of alum in baking-powders might have been presented, but it would have been of a similar nature to that which has already been given. In the writer's opinion, the presence of alum in baking-powders is objectionable, since, under certain conditions, it may exert an injurious effect on the digestion. The effects may not be very marked in the case of any individual consumer, but that they can be induced to a greater or less extent seems to be well established.

There appears to be ample ground for requiring that the makers of baking-powders should publish the ingredients used in their powders, in order that the consumer, who may justly have doubts of the desirability of using certain kinds, may be protected. At present the only guaranty of an undoubtedly wholesome and efficient article appears to be the name of the brand.

Moreover, since it is quite possible to put up the baking-powders in such a way as to preserve their strength very thoroughly, and since it is evident that many makers fail in this respect, it would not seem unreasonable to require that baking-powders should not be sold unless they will yield a certain percentage of carbonic-acid gas. The bad effects of the "heavy" food prepared with some of the baking-powders among our samples must certainly be felt by those who use them, and who are yet too ignorant to know where the trouble lies. It is for this class especially that nearly all legislation relating to securing good food and drugs is enacted.

Since it is evident that some of the alum powders are so prepared as to increase the extent of any injurious effect, owing to the mixture of ingredients whose combination can not be justified on any grounds, it is recommended that a special and more thorough examination of such be made, with a view to preventing their manufacture.

APPENDIX.

1. In bread made from Orange baking-powder, page —, according to directions, there was found, in a condition readily soluble in tepid water, alumina equivalent to 26 grains of crystallized ammonia alum per 2-pound loaf.

2. With reference to Professor Patrick's experiments on cats, the writer had biscuit made with a "straight" alum baking-powder, using twelve times the proper amount of the powder. The biscuit had a bitter, nauseating taste, and must have been very indigestible, so that no fair conclusions could be drawn from its use.

ANALYSES BY U. S. DEPARTMENT OF AGRICULTURE.

The samples for the following analyses were all purchased at retail stores in Washington, D. C., by an agent of the Department, no intimation being given to the seller of the purpose for which they were intended. The city was pretty thoroughly gone over, and the samples probably include about all the different brands sold.

METHODS OF ANALYSIS.

Following are detailed descriptions of the methods followed in the analyses made in this laboratory. In many of the estimations different methods were tried, and the one which gave the best results and was found to be most convenient was chosen.

GENERAL EXAMINATION OF SAMPLES.

The qualitative and general examination is described in the extract from Professor Cornwall's report above, and it is not necessary to go into it in detail again, as the methods were similar, generally speaking. The qualitative examination and assignment of a sample to one of the classes indicated presents no special difficulties. If it is desired to know the character of the filling used, it is readily ascertained by a microscopical examination; but this is rather an unimportant matter. A determination of the alkalimetry of the watery solution of the powder is useful as showing whether any great excess of alkali had been used.

ESTIMATION OF CARBONIC ACID.

This is one of the most important estimations, as it determines the strength of the powder. It was made by absorption with soda lime, and a form of apparatus was used that has served for some time in this laboratory for the determination of carbonic acid. This apparatus has recently been somewhat modified and greatly improved in compactness by Mr. A. E. Knorr. It is shown in the accompanying figure. Following is Mr. Knorr's description: [1]

The apparatus proper, as represented by the cut, consists of a flask A in which the carbon dioxide is set free. A condenser D is ground into the neck of this flask and condenses the steam formed when the liquid in A is boiled in order to secure complete expulsion of the gas. The reservoir B contains the acid required for the operation, and has a soda-lime guard C ground into it to retain the carbonic acid of the air, a constant current of which is aspirated through G during the whole operation. The stem of the reservoir is ground into the condenser, or it may be conveniently blown into one piece with it. The carbonic acid is dried in E and finally absorbed in the weighed potash bulb F.

Two determinations of carbonic acid were made on each sample—one by the addition of acid to determine the total amount of carbonate present, and one by the addition of water only. The per cent. of carbonic acid found in the latter estimation may properly be termed the *available* amount present in the powder, as it is the quantity which would be actually liberated by the acid ingredient of the powder when it is used in baking, and therefore represents the actual *value* of the powder for aerating purposes, so far as the *evolution* of gas is concerned.

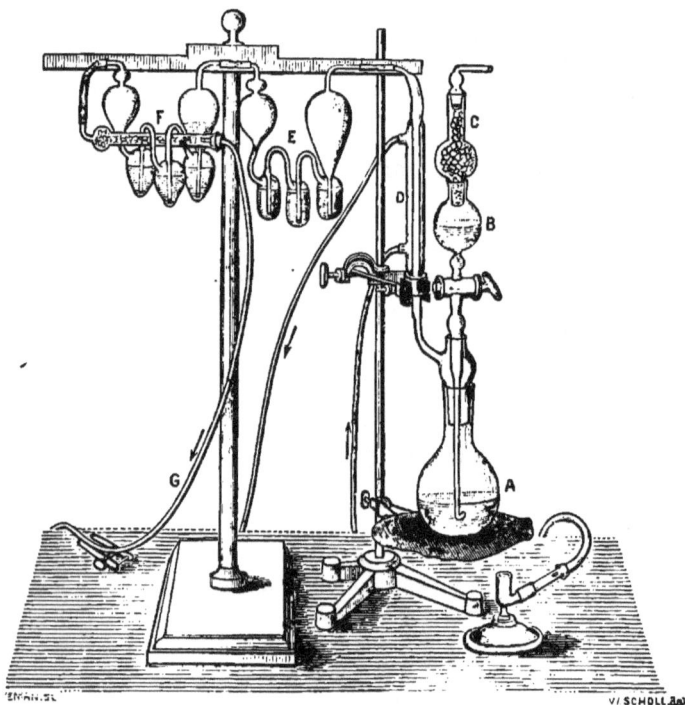

FIG. 26.

For the determination of the *total* CO_2 the procedure was as follows: Place in a short glass tube, the weight of which is known, about 1 to 2 grams of the sample, and weigh the whole as quickly as possible, the amount taken being obtained by difference. The tube and contents are gently dropped into the generating flask of the apparatus, which must be perfectly dry. The flask is closed with the stopper carrying the tube connecting with the absorption apparatus, and also the funnel tube, which has been previously provided with 10 cubic centimeters of dilute sulphuric acid for the liberation of the gas. When all parts of the apparatus are connected, and seen to be tight, the stopper of the funnel tube is opened, and the acid allowed to run slowly into the flask. The generation of the gas should be made as gradual as possible; by running in a small quantity of the acid at first and tilting the flask slightly this can readily be accomplished; after the greater part of the sample has been acted upon by the acid and before the latter has all been added, a lamp is placed under it, and the contents gradually heated to boiling, gentle aspiration being made at the same time. The operation is finally finished by the funnel tube being opened, and air, free from CO_2, drawn through it and through the apparatus, the contents of the flask at the same time being kept at ebullition. This is continued for fifteen minutes, when the absorption tube is removed from the apparatus, allowed to cool, and weighed. Its increase in weight gives the amount of CO_2 absorbed. The determination of the *available* CO_2 was conducted in a similar manner, with the substitution of pure boiled water in the funnel tube instead of acid. After the sample has all been acted on, the contents of the flask are just brought to a boil, then the lamp is removed and air is drawn through the liquid for exactly fifteen minutes. The conditions were made as nearly alike as possible for each sample in this estimation, for, different results can readily be obtained by varying them. The above conditions are believed to be as close an approximation to those actually obtaining in the use of the powder as can be arrived at in an ordinary chemical analysis. Prolonged boiling of the liquid residue is inadvisable, for in case the ingredients in the powder are not accurately proportioned, and a considerable excess of bicarbonate is present, long boiling will liberate gas from it after the acid ingredient has all been neutralized, and thus a high result is obtained from a poorly-made sample, while in one where the bicarbonate is not greatly in excess of the proper amount, the above procedure will readily give the full amount available. *

*In some experiments made to determine the amount of carbonic acid driven off from bicarbonate of soda on heating its water solution, the following results were obtained: (1) Just brought to a boil under the same conditions as in the determinations made above, 6.99 per cent. of the weight of the bicarbonate was obtained ; (2) Boiled 15 minutes, 16.17 per cent. was obtained ; and, (3) boiled 1½ hours, 20.70 per cent., or about the full quantity of acid carbonate.

ESTIMATION OF STARCH.

This estimation was made by the well-known method of conversion by heating with acid and the determination of the copper oxide reducing power of the resultant solution. While by no means satisfactory, this is probably the best method we have at present for starch estimation. No difficulty was found in its application to all classes of baking-powders, the other ingredients offering no obstacle to its proper performance. To insure agreeing results it is very essential to conduct the conversion under precisely the same conditions in all cases.

The following is the detailed procedure :

From 2 to 5 grams are weighed out and transferred to an Erlenmeyer flask; to it are added about 150 to 200 cubic centimeters of a solution of hydrochloric acid which has a strength of 4 per cent. of the acid gas. The flask is provided with a cork, perforated for the passage of a condensing tube about 1 meter in length. The conversion is accomplished by gently boiling the acid liquid for four hours, after which the flask and contents are cooled, neutralized by the addition of sodium hydrate, made up to a definite volume and the copper oxide reducing power determined. The latter operation is best carried out by the method used in this laboratory, in which asbestos-tipped filtering tubes are used for the end reaction.[1] The reducing power being calculated as dextrose; 10 parts equal 9 parts of anhydrous starch.

Professor Weber used a rough method for the direct estimation of starch in his samples, which he describes as follows:[2]

One gram was weighed off, transferred to a small beaker, covered with water, allowed to stand until action had ceased, filtered and washed, residue spurted by means of a wash-bottle into a flat-bottomed platinum dish, allowed to settle, the supernatant water removed as far as possible by means of a pipette, the remainder of water evaporated, the residue dried at 100° C. and weighed. The residue was then incinerated and the amount of ash deducted from above weight. In case of alum powders the ash remaining after ignition was Al_2O_3, which was contained in the residue dried at 100° C. as $Al_2(OH)_6$; consequently the Al_2O_3 was calculated as $Al_2(OH)_6$ before deducting.

This method was carried out upon the entire series of samples examined here. In many cases it gave results agreeing quite closely with those obtained by the direct estimation, but in some samples the results were entirely too high. It is not applicable to alum powders even with the correction made above. For a rough method it answers fairly well and it is quite easy of execution.

ESTIMATION OF PHOSPHORIC ACID.

This determination was made in the same manner as in fertilizers, as prescribed by the Association of Official Agricultural Chemists at their last meeting, as follows:[3]

Weigh out 2 grams of the sample, ignite carefully in a muffle, and treat with 30 cubic centimeters concentrated nitric acid.

Boil gently until all phosphates are dissolved and all organic matter destroyed; dilute to 200 cubic centimeters; mix and pass through a dry filter; take 50 cubic centimeters of filtrate; neutralize with ammonia. To the hot solutions for every decigram of P_2O_5 that is present add 50 cubic centimeters of molybdic solution. Digest at about 65° C. for one hour, filter, and wash with water or ammonium nitrate solution. (Test the filtrate by renewed digestion and addition of more molybdic solution.) Dissolve the precipitate on the filter with ammonia and hot water and wash into a beaker to a bulk of not more than 100 cubic centimeters. Nearly neutralize with hydrochloric acid, cool, and add magnesia mixture from a burette; add slowly (one drop per second), stirring vigorously. After fifteen minutes add 30 cubic centimeters of ammonia solution of density 0.95. Let stand several hours (two hours

---

[1] Fully described in Bull. No. 15, p. 32.
[2] Communicated to the author in MSS.
[3] Bull. No. 19, Chem. Div. U. S. Dep't Ag'l, p. 58.

are usually enough). Filter, wash with dilute ammonia, ignite intensely for ten minutes, and weigh.

ESTIMATION OF TARTARIC ACID.

The method used in this estimation was that known as the " Goldenberg Geromont," which is described in full in *Chemiker Zeitung*, 12, 1888, 390. This and other methods for the estimation of tartaric acid in crude argol and other raw materials were lately submitted to a critical comparison in the tartrate factory at Nienburg;[1] the Goldenberg-Geromont method avoided the principal sources of error and is recommended as the best and most easy of execution.

The procedure as modified for application to a tartrate baking-powder is as follows:

Weigh out 5 grams, wash into a 500-cubic centimeter flask with about 100 cubic centimeters of water; add 15 cubic centimeters of strong hydrochloric acid; make up to mark and allow the starch to settle. Filter, measure out 50 cubic centimeters of the clear filtrate; add to it 10 cubic centimeters of a solution of carbonate of potash containing 300 grams K_2CO_3 to the liter and boil half an hour; filter into a porcelain dish and evaporate filtrate and washings to a bulk of about 10 cubic centimeters. Add gradually with constant stirring 4 cubic centimeters glacial acetic acid, and then 100 cubic centimeters of 95 per cent. alcohol, stirring the liquid until the precipitate floating in it assumes a crystalline appearance. After it has stood long enough for this precipitate to form and settle, best for several hours, decant through a small filter, add alcohol to the precipitate, bring it on the filter, wash out the dish and finally the filter carefully, with alcohol, until it is entirely free from acetic acid. Transfer filter and precipitate to a beaker, add water and boil, washing out the dish also with boiling water if any of the precipitate adheres to it. The resulting solution is titrated with decinormal alkali solution, using phenol-phthalein as indicator; 1 cubic centimeter decinormal alkali corresponds to .0188 grams of potassium bitartrate, or .0150 grams of tartaric acid.

ESTIMATION OF ALKALIES.

The estimation of the soda and potash in the powders was carried out by separating them as chlorides, determining the potash as potassium platinic chloride and calculating the difference as sodium chloride. The detailed procedure was similar to that used by the Association of Official Agricultural Chemists for determining potash in fertilizers, as follows:

Weigh out 5 grams into a platinum dish and incinerate in a muffle at a low heat. The charred mass is well rubbed up in a mortar, then boiled 15 minutes with about 200 cubic centimeters of water to which

[1] Chemiker Zeitung 13, 1889, 160.

has been added a little hydrochloric acid. The whole is transferred to a 500-cubic centimeter flask and after cooling made up to the mark and filtered. Of the filtered liquid 100 cubic centimeters, representing 1 gram of the sample, are measured out, heated in the water bath, and slight excess of barium chloride added; then without filtering barium hydrate is added in slight excess, the precipitate filtered off and washed. To the filtrate is added a little ammonium hydrate and then ammonium carbonate until all the barium is precipitated. This precipitate is filtered and washed, the filtrate evaporated to dryness and carefully ignited until all volatile matter is driven off, when it is weighed. This gives the weight of the mixed chlorides. The residue is taken up with hot water, from 5 to 10 cubic centimeters of a 10 per cent. solution of platinic chloride added, and the whole evaporated to a sirupy consistence in the water bath ; then it is treated with strong alcohol. the precipitate washed with alcohol by decantation, transferred to a Gooch crucible, dried at 100° C., and weighed. The weight of the precipitate multiplied by .19308 gives the weight of K_2O, and by .3056 the equivalent amount of KCl. The weight of KCl found is subtracted from the weight of the mixed chloride, the remainder being the NaCl, which multiplied by .5300 gives the weight of Na_2O in the sample.

ESTIMATION OF ALUMINIUM.

In the case of a " straight " alum powder, the simple estimation by burning to ash, extracting, and determining the alum by direct precipitation with ammonia would probably be accurate, but in view of the frequent use of flour as a " filler," as well as of the presence of calcium as an impurity, it is best, even with those made up with alum alone, to use a method which will insure a complete separation of the alum. The following procedure, given by Allen for the quantitative estimation of alum in bread, was found to give good results with baking-powders :

Take 5 grams and incinerate in a muffle. The heat should be moderate so as not to fuse the ash. The process is completed by adding pure sodium carbonate and a little niter, and heating the mixture to fusion. The product is rinsed out with water into a beaker, acidulated with hydrochloric acid, and evaporated to dryness. The residue is taken up with dilute acid, the liquid made up to 500 cubic centimeters in a graduated flask, filtered, and 50 cubic centimeters taken for precipitation. To the solution dilute ammonia is added until the precipitate barely redissolves on stirring, when a slightly acid solution of ammonium acetate is added, and the liquid brought to a boil. After a few minutes' heating the solution should be set aside for some hours, when its appearance should be observed. (If gelatinous it probably consists solely of iron and alum phosphates, but if granular more or less of the earthy phosphates have been co-precipitated; then it should be separated, redissolved in dilute hydrochloric acid, the solution again neutralized with ammonia, and treated with ammonium acetate.) The precipitate of iron and aluminium phosphates is filtered off, washed, and redissolved in the smallest quantity of hydrochloric acid. The resultant solution is poured into an excess of an aqueous solution of *pure* caustic soda contained in a platinum or nickel vessel. After heating for some time, the liquid is considerably diluted and filtered. The filtrate is acidulated with hydrochloric acid, ammonium acetate and a few drops of sodium phosphate added, and then a slight excess of ammonia. The liquid is kept hot till all smell of ammonia is lost, when it is filtered, and the precipitated aluminium phosphate washed, ignited, and weighed. Its weight multiplied by 3.713 gives the ammonia alum (hydrated), or by 3.873 the potash alum in the .5 grams of sample taken.

In the phosphate and alum powders the above method gave a fairly good separation of the alum, but the following separation by means of molybdenum was found to be more exact, and at the same time much more convenient of application. It was adapted to the powders by Mr. K. P. McElroy.

Weigh out 5 grams into a platinum dish, char, treat with strong nitric acid, and filter into a 500 cubic centimeter flask. After washing the residue slightly, transfer filter and all back to the platinum dish and burn to whiteness. To the ash add mixed carbonates and fuse. Take up with nitric acid, evaporate to dryness, acidify again with nitric acid, and wash all into the 500 cubic centimeter flask. Nearly neutralize the contents of the flask with ammonia, and add molybdate of ammonium sufficient to precipitate all the phosphoric acid present. Allow to stand some time, make up to the mark, shake thoroughly, and filter off 100 cubic centimeters through a dry filter. This is exactly neutralized with ammonia, keeping the solution as cool as possible to avoid the deposition of molybdic acid. Filter and wash the precipitate, redissolve in dilute nitric acid with the addition of a little ammonium nitrate, and precipitate as before. Filter through a paper filter, burn, ignite, and weigh as Al_2O_3. The alumina and phosphoric acid may be determined in the same sample by the above method, modifying it as follows: When the solution, ash, etc., have all been brought into the graduated flask, make up to the mark without adding molybdate, filter and take 100 cubic centimeters, nearly neutralize with ammonia, add ammonium nitrate and molybdate of ammonium, digest and filter. The filtrate contains the aluminium and may be precipitated with ammonia as above, while the phosphoric acid is all contained in the precipitate, which may be redissolved in ammonia and precipitated with magnesia mixture.

ESTIMATION OF CALCIUM.

This determination was made as follows: Weigh out 5 grams of the sample, transfer to a 500 cubic centimeter flask, add 40 or 50 cubic centimeters of water, and then 20 or 30 cubic centimeters of strong hydrochloric acid. Make up to the mark with water, shake thoroughly, and set aside to allow starch to settle. Filter through a dry filter, and take aliquot parts of the filtrate for precipitation ; in phosphate powders not more than 50 cubic centimeters should be used. Nearly neutralize with ammonia, acidify slightly with acetic acid, add ammonia acetate, and boil. Filter from the precipitate, if there be any, add ammonium oxalate, and allow to stand several hours. Filter into a Gooch crucible, and dry at 100°. Weigh as oxalate.

ESTIMATION OF SULPHURIC ACID.

The sulphuric acid was estimated without previous ignition of the powder, as follows :

Weigh out .5 to 1 gram of the powder, according to its character, the former quantity being more convenient for alum powders, and transfer to a beaker. Digest with strong hydrochloric acid until all of the powder, including the starch, goes into solution ; add barium chloride to slight excess while still hot, and allow it to stand for twelve hours, or over night. Filter through a Gooch crucible, ignite, and weigh.

ESTIMATION OF AMMONIA.

Ammonia is present either as bicarbonate, or as ammonium sulphate in the alum powders. The estimation was made by the Kjeldahl method, as used by the Association of Official Agricultural Chemists.[1] Where flour instead of starch is used as a filling the gluten would give ammonia, of course, and wherever a tartrate powder was found to give any appreciable amount of ammonia by the method, a weighed portion was taken, water added, the solution filtered from the starch, and subjected to analysis. The results were practically the same as those obtained directly from the powder. Probably flour is not often used. In the case of the alum powders, the difference that would be made by flour filling was disregarded, as the amount of alum present is sufficiently established by the percentage of aluminium oxide and sulphuric acid found ; the amount of ammonia found was almost invariably low in proportion to these other constituents of ammonia alum.

ESTIMATION OF MOISTURE.

The percentage of water of association and combination as given in the analyses was obtained by difference. A number of attempts were made to estimate it directly in the following way : A weighed portion of the sample was placed in a U tube, which was kept immersed in boiling water. At one end this tube was connected with a series of sulphuric-acid wash-bottles, and at the other with weighed potash bulbs filled with sulphuric acid, and beyond these with an aspirator. In this way a current of dried air was drawn through the sample while it was kept at 100° C., and the water drawn from it in this way was absorbed by the sulphuric acid in the potash bulb, while the carbonic acid was drawn

---

[1] Fully described in Bull. No. 19, Chem. Div. U. S. Dep't Agriculture.

into the aspirator. The increase in weight of the potash bulbs gave the weight of water absorbed. It was found, however, that the sample would cake into a hard mass, through which a channel would form which would permit the passage of the current of dry air, without affecting the greater mass of the powder, and no exact results could be obtained. Some improvement was made by mixing the powder with dry oxide of zinc, so as to prevent the formation of a channel, but still the results were not at all satisfactory, and the attempt to make a direct estimation was finally abandoned. Even if the determination could be made exact, it is doubtful if all the water of combination could be obtained at 100° C., especially in phosphate and alum powders, and probably a temperature high enough to accomplish this would effect a decomposition of the starch.

EXPRESSION OF THE RESULTS OF ANALYSIS.

The results of analysis are given, first, as acid and basic radicals in percentage composition while in the second part of each table an attempt has been made to combine these into salts showing the constitution of the powder. The difficulties attending this calculation of the probable combination of the acids and bases were so great that I was frequently tempted to give it up entirely and state only percentage composition. I finally concluded to insert the calculation with the proviso that it should be considered at best merely an approximation to the exact composition of the powder. The obstacles in the way of an exact calculation may be stated as follows: In the first place the amount of total carbonic acid found is always less than that required to form bicarbonate of soda with the amount of sodium oxide found. This is undoubtedly due to a partial action of the acid constituent of the powder upon the bicarbonate, by which carbonic acid has been lost. The percentage of bicarbonate is therefore calculated from the per cent. of carbonic acid found, and the excess of sodium oxide left is stated as "residual" sodium oxide, without attempting to make further hypotheses as to the results of its combination with the acid constituent. It is possible that part of the bicarbonate may become converted into the normal carbonate under the conditions of being mixed and in contact with other chemicals, though this is not likely; then the indefinite composition of many of the commercial salts used in the powders renders it an extremely difficult matter to arrive at any satisfactory conclusion as to the make up of the powder in which they are used. The acid phosphate of lime, for instance, is a very variable substance, and even ammonia alum, which might reasonably be supposed to be constant in its composition, is found to vary widely from the theoretical. Its content of water varies according to the greater or less amount of drying it has undergone, and aside from this the ratios of the ammonia, sulphuric acid, and alumina to one another are at variance with the formula.

This is shown by the following analysis of a sample of commercial ammonia alum obtained in a powdered condition :

Analysis of commercial ammonia alum.

Constituents.	Found.	Theoretical composition.
	Per cent.	*Per cent*
Aluminium oxide, Al_2O_3	12. 62	11. 36
Sulphuric acid, SO_3	34. 17	35. 28
Ammonia, NH_3	2. 75	3. 75
Water of crystallization (by difference)	50. 46	49. 61
	100. 00	100. 00

This indicates that the commercial salt is somewhat basic as regards the alumina, yet there is a deficiency of ammonia, so that if the former is all combined with sulphuric acid as normal sulphate, there is still not sufficient of the acid left for combination with the ammonia, although the latter is present in too low a proportion to the other constituents. This anomaly holds good throughout many of the samples.

I have expressed the sum total of the alumina, sulphuric acid, and ammonia as "anhydrous ammonia alum," combining the sulphuric acid, first with the alumina as far as it went, and the rest with the ammonia, and where there was not sufficient for combination with all the ammonia adding the latter *as ammonia.*

The presence of acid phosphate of lime still further complicates this calculation, as it is a question how much of the sulphuric acid should be taken from the alum to combine with part of the lime as sulphate of lime.

In the expression of the results where acid phosphate of lime is present, I have combined the lime with phosphoric acid as normal phosphate as far as it went, and added the rest as free phosphoric acid, grouping the whole together and calling it "acid phosphate of lime." Following is an analysis of a sample of commercial acid phosphate of lime, obtained from the trade :

Analysis of commercial acid phosphate of lime.

	Per cent.
Calcium oxide, CaO	24.93
Phosphoric acid, P_2O_5	52.45
Sulphuric acid, SO_3	.15
Water	22.80
	100.33

In this sample the ratio of lime to phosphoric acid is about 1 : 2, and this relation holds good in many of the powders containing the phosphate, but in some it is quite different. The above sample is almost free from sulphate of lime, while many of the powders show considerable quantities of it, indicating that all the acid phosphates are not so pure in this respect. Chemists will readily understand the impossibility of giving the proportions of the various forms of lime phosphates contained in such a substance. As given in the tables the relative acidity is shown, though of course the phosphoric acid does not occur as free acid.

Both the alum and the lime phosphate contain large percentages of water, hygroscopic moisture in the latter, and crystallization water in the former, so that the per cents. of the "anhydrous" salts given are always lower than the proportions of the hydrated salts originally used in compounding the powder. Nearly half the weight of the alum is crystallization water, some part of which is probably driven off in some cases when it is used for baking-powder purposes, but there are no means of ascertaining how much, and of course the moisture in the acid phosphate would vary in different samples, so there is no possible way of approximating the amounts of the hydrated substances, as they were originally used.

Ammonia bicarbonate is another substance of indefinite composition. As given in the tables it has been calculated from the ammonia found, upon the assumed composition given it in the U. S. Pharmacopœia, viz, $NH_4HCO_3.NH_4NH_2CO_2$.

In case ammonia carbonate were present in any of the powders containing ammonia alum, I know of no way of estimating the amount or even the fact of its presence in the small quantities used.

The percentage of "available carbonic acid" is placed first in the tables as constituting the most important indication of the efficiency of the sample as an aerating agent.

TARTRATE POWDERS.

5503.—*Roya Baking-Powder.*

[Manufactured by The Royal Baking-Powder Company, New York]

Available carbonic acid ....per cent..	**12.74**
Cubic inches per ounce of powder at 212° F....................................	153.0
Leavening gas (available $CO_2 + NH_3$) ....per cent..	13.06
Cubic inches per ounce of powder at 212° F...................................	160.6

PERCENTAGE COMPOSITION.

Total carbonic acid, CO_2	12.92
Sodium oxide, Na_2O	10.30
Potassium oxide, K_2O..	12.02
Calcium oxide, CaO..	.13
Ammonia, NH_3..	.32
Tartaric acid, $C_4H_4O_5$...	37.46
Sulphuric acid, SO_3...	.25
Starch...	16.34
Water of combination and association by difference.....................	10.26
	100.00

PROBABLE PERCENTAGE COMBINATION.

Sodium bicarbonate, $NaHCO_3$	23.61
(Residual sodium oxide. Na_2O)	1.59
Ammonium bicarbonate, $N_3H_{11}C_2O_5$	.98
Potassium bitartrate, $KHC_4H_4O_6$	53.34
Calcium sulphate, $CaSO_4$	.31
Starch	16.34
Water of association	3.83
	100.00

This powder contains a small quantity of ammonium bicarbonate.

5504.—Dr. Price's Cream Baking Powder.

[Made by Price Baking-Powder Company, New York and Chicago.]

Available carbonic acid ....per cent..	**11.13**
Cubic inches per ounce of powder at $212°$ F	133.6

PERCENTAGE COMPOSITION.

Total carbonic acid, CO_2	12.25
Sodium oxide, Na_2O	11.03
Potassium oxide, K_2O	11.71
Calcium oxide, CaO	.19
Tartaric acid, $C_4H_4O_5$	35.14
Sulphuric acid, SO_3	.12
Starch	18.43
Water of combination and association by difference	11.13
	100.00

PROBABLE PERCENTAGE COMBINATION.

Sodium bicarbonate, $NaHCO_3$	23.38
(Residual sodium oxide, Na_2O)	2.40
Potassium bitartrate, $KHC_4H_4O_6$	50.04
Calcium sulphate, $CaSO_4$	.20
Starch	18.43
Water of association	5.55
	100.00

5505.—Cleveland's Superior Baking-Powder.

[Made by Cleveland Brothers, 911 and 913 Broadway, Albany, N. Y.]

Available carbonic acid ....per cent..	**12.58**
Cubic inches per ounce of powder at $212°$ F	151.1

PERCENTAGE COMPOSITION.

Total carbonic acid, CO_2	13.21
Sodium oxide, Na_2O	13.58
Potassium oxide, K_2O	14.93
Calcium oxide, CaO	.18
Tartaric acid, $C_4H_4O_5$	41.60
Sulphuric acid, SO_3	.10
Starch	7.42
Water of combination and association by difference	8.98
	100.00

Sodium bicarbonate, $NaHCO_3$	25.21
(Residual sodium oxide, Na_2O)	4.28
Potassium bitartrate, $KHC_4H_4O_6$	59.25
Calcium sulphate, $CaSO_4$	.17
Starch	7.42
Water of association	3.67
	100.00

5507.—Sea Foam (Gantz) Baking-Powder.

[Made by Gantz, Jones & Co., 176 Duane street, New York.]

Available carbonic acid ....per cent..	**8.03**
Cubic inches per ounce of powder at 212 ° F	96.5

PERCENTAGE COMPOSITION.

Total carbonic acid, CO_2	8.10
Sodium oxide, Na_2O	15.47
Potassium oxide, K_2O	15.24
Calcium oxide, CaO	.78
Tartaric acid, $C_4H_4O_6$	44.18
Starch	5.32
Water of combination and association by difference	10.91
	100.00

PROBABLE PERCENTAGE COMBINATION.

Sodium bicarbonate, $NaHCO_3$	15.46
(Residual sodium oxide, Na_2O)	9.77
Potassium bitartrate, $KHC_4H_4O_6$	62.92
(Calcium oxide, CaO)	.78
Starch	5.32
Water of association	5.75
	100.00

This and the preceding, No. 5505, contain very small quantities of starch, apparently rather too little for the proper preservation of the last sample.

5522.—Hecker's Perfect Baking-Powder.

[Made by George V. Hecker & Co., 205 Cherry street, New York.]

Available carbonic acid	**9.29**
..........per cent..	
Cubic inches per ounce of powder at 212° F	111.6

PERCENTAGE COMPOSITION.

Total carbonic acid, CO_2	9.26
Sodium oxide, Na_2O	11.61
Potassium oxide, K_2O	11.63
Calcium oxide, CaO	.91
Tartaric acid, $C_4H_4O_5$	39.74
Sulphuric acid, SO_3	.22
Starch	12.78
Water of combination and association by difference	13.85
	100.00

PROBABLE PERCENTAGE COMBINATION.

Sodium bicarbonate, $NaHCO_3$	17. 67
(Residual sodium oxide, Na_2O)	5. 09
Potassium bitartrate, $KHC_4H_4O_6$	56. 60
Calcium sulphate, $CaSO_4$	.37
(Calcium oxide, CaO)	.76
Starch	12. 78
Water of association	6. 73
	100. 00

This sample contains rather more lime than most of the other tartrate powders. The excess given above is probably combined with tartaric acid as calcium tartrate; the same is true of the two following samples:

5527.—*Gilbert S. Graves's Imperial Baking-Powder.*

[Made by The Imperial Baking-Powder Company, Buffalo, N. Y.]

Available carbonic acid	**7.28**
Cubic inches per ounce of powder at 212° F	87. 4

PERCENTAGE COMPOSITION.

Total carbonic acid, CO_2	8. 47
Sodium oxide, Na_2O	13. 62
Potassium oxide, K_2O	9. 42
Calcium oxide, CaO	.45
Tartaric acid, $C_4H_4O_5$	32. 74
Starch	24. 57
Water of combination and association by difference	10. 73
	100. 00

PROBABLE PERCENTAGE COMBINATION.

Sodium bicarbonate, $NaHCO_3$	16. 16
(Residual sodium oxide, Na_2O)	7. 66
Potassium bitartrate, $KHC_4H_4O_6$	46. 63
(Calcium oxide, CaO)	.45
Starch	24. 57
Water of association	4. 53
	100. 00

5529.—*Thurber's Best Baking-Powder.*

[Made by H. K. & F. B. Thurber & Co., West Broadway, Reade and Hudson streets, New York.]

Available carbonic acid per cent..	**10 26**
Cubic inches per ounce of powder at 212° F	123. 2

PERCENTAGE COMPOSITION.

Total carbonic acid, CO_2	10. 54
Sodium oxide, Na_2O	10. 65
Potassium oxide, K_2O	12. 26
Calcium oxide, CaO	.66
Tartaric acid, $C_4H_4O_5$	38. 75
Sulphuric acid, SO_3	.07
Starch	13. 41
Water of combination and association by difference	13. 66
	100. 00

Sodium bicarbonate, NaHCO₃	20.12
(Residual sodium oxide, Na₂O)	3.22
Potassium bitartrate, KHC₄H₄O₆	55.19
Calcium sulphate, CaSO₄	.12
(Calcium oxide, CaO)	.61
Starch	13.41
Water of association	7.33
	100.00

5513.—Sterling Baking-Powder.

[Made by Sterling Manufacturing Company, Baltimore, Md.]

Available carbonic acid ....per cent..	**9.53**
Cubic inches per ounce of powder at 212° F	114
Leavening gas (available $CO_2 + NH_3$) ....per cent..	9.90
Cubic inches per ounce of powder at 212° F	123.3

PERCENTAGE COMPOSITION.

Total carbonic acid, CO_2	10.66
Sodium oxide, Na₂O	10.38
Potassium oxide, K₂O	.66
Calcium oxide, CaO	.15
Ammonia, NH₃	.37
Tartaric acid, C₄H₄O₅	21.94
Starch	40.05
Water of combination and association by difference	15.79
	100.00

PROBABLE PERCENTAGE COMBINATION.

Sodium bicarbonate, NaHCO₃	19.13
(Residual sodium oxide, Na₂O)	3.32
Ammonium bicarbonate, N₃H₁₁C₂O₅	1.14
Free tartaric acid, H₂C₄H₄O₆	24.93
Starch	40.05
Water of association	11.43
	100.00

This and the following are the only samples examined which were made up with *free* tartaric acid as the sole acid constituent. Both contain small quantities of ammonium bicarbonate. The small quantities of potassium and calcium oxide probably exist as tartrates.

5535.—Our Best Baking-Powder.

[Made by the Purity Chemical Works, Philadelphia, Pa.]

Available carbonic acid ....per cent..	**4.98**
Cubic inches per ounce of powder at 212° F	9.8
Leavening gas (available $CO_2 + NH_3$) ....per cent..	5.22
Cubic inches per ounce of powder at 212° F	65.5

PERCENTAGE COMPOSITION.

Total carbonic acid, CO_2	5.13
Sodium oxide, Na₂O	13.65
Calcium oxide, CaO	.07
Ammonia, NH₃	.24
Tartaric acid, C₄H₄O₅	18.48
Sulphuric acid, SO₃	.61
Starch	45.63
Water of combination and association by difference	16.19
	100.00

PROBABLE PERCENTAGE COMBINATION.

Sodium bicarbonate, $NaHCO_3$	9.01
(Residual sodium oxide, Na_2O)	10.32
Ammonium bicarbonate, $N_3H_{11}C_2O_5$	.74
Free tartaric acid, $H_2C_4H_4O_6$	21.00
(Sulphuric acid, SO_3)	.61
Starch	45.63
Water	12.69
	100.00

The sulphuric acid in this sample probably is present as sodium sulphate, which may be simply an impurity in some of the salts.

PHOSPHATE POWDERS.

5506.—*Wheat Baking-Powder.*

[Made by Martin Kalbfleisch's Sons, New York.]

Available carbonic acid ....per cent..	**3.79**
Cubic inches per ounce of powder at $212°$ F	45.5

PERCENTAGE COMPOSITION.

Total carbonic acid, CO_2	5.57
Sodium oxide, Na_2O	14.10
Potassium oxide, K_2O	7.49
Calcium oxide, CaO	11.96
Phosphoric acid, P_2O_5	36.69
Sulphuric acid, SO_3	.16
Ammonia, NH_3	.16
Water of combination and association by difference	23.87
	100.00

This powder is quite peculiar in its make-up. It contains no starch or filling and has a very low per cent. of available gas. It contains a small quantity of ammonia bicarbonate. It contains a considerable amount of potassium oxide, which would indicate the presence of an acid phosphate of potash, as there is no other possible combination for this base, and the per cent. of lime is hardly sufficient to answer for all the phosphoric acid present.

5508.—*Rumford Yeast Powder.*

[Made by Rumford Chemical Works, Providence, R. I.]

Available carbonic acid ....per cent..	**12.86**
Cubic inches per ounce of powder at $212°$ F	154.5

PERCENTAGE COMPOSITION.

Total carbonic acid, CO_2	13.17
Sodium oxide, Na_2O	12.66
Potassium oxide, K_2O	.31
Calcium oxide, CaO	10.27
Phosphoric acid, P_2O_5	21.83
Starch	26.41
Water of combination and association by difference	15.05
	100.00

PROBABLE PERCENTAGE COMBINATION.

Sodium bicarbonate, NaHCO₃		25.71
(Residual sodium oxide, Na₂O)		3.17
Acid phosphate of lime, anhydrous:		
Ca₃(PO₄)₂	18.95	
H₃PO₄	18.15	
		37.10
Starch		26.41
Water of association (phosphate)		7.61
		100.00

5509. —*Horsford's Self-raising Bread Preparation.*

[Made by Rumford Chemical Works, Providence, R. I.]

THE ACID POWDER.

PERCENTAGE COMPOSITION.

Calcium oxide, CaO	16.78
Phosphoric acid, P₂O₅	23.97
Sulphuric acid, SO₃	6.00
Starch	20.81
Water of combination and association by difference	32.44
	100.00

PROBABLE PERCENTAGE COMBINATION.

Acid phosphate of lime:		
Ca₃(PO₄)₂	21.00	
H₃PO₄	19.80	
		40.80
Calcium sulphate, CaSO₄		11.40
Starch		20.81
Water of association		26.99
		100.00

This sample was different from all the others in being put up in two separate packages. One of these contained the acid ingredient, with starch to keep it from becoming dry, and the other, bicarbonate of soda. The directions are to mix the contents of both papers, if the whole is to be made use of at once, or to use two equal measures of the acid part to one of the soda. For analysis the contents of the papers were taken separately, and the results obtained from the acid part are given above. The other paper contained simply bicarbonate of soda of good strength.

For the determination of the available carbonic acid another sample was purchased, the entire contents of both papers thoroughly mixed, and a portion of the mixed powders weighed out and submitted to analysis. The first estimation made gave 13.56 per cent. available carbonic acid. By the time a duplicate estimation was made, perhaps an hour after the first, the per cent. had fallen to 12.03, showing a very rapid deterioration, or loss of available gas. Two subsequent estimations made the same day gave, respectively, 9.50 per cent. and 9. per cent. Two determinations made the next day gave 8.35 per cent. and 8.35 per cent. This plan of keeping the acid and alkali ingredients separate appears to be an excellent one, though it is similar to the old method of using cream of tartar and baking-soda, with acid phosphate of lime substituted for the cream of tartar. The acid phosphate used in this sample contains considerable sulphate. The preceding sample (No. 5508), though made by the same firm, is entirely free from it, showing that it had been made from a better article of phosphate.

ALUM POWDERS.

5526.—*Vienna Baking-Powder.*

[Made by the Penn Chemical Works, Philadelphia, Pa.]

Available carbonic acid ....per cent..	**6.41**
Cubic inches per ounce of powder at 212° F................................	77. 0

PERCENTAGE COMPOSITION.

Total carbonic acid, CO_2	7. 90
Sodium oxide, Na_2O	6. 99
Calcium oxide, CaO..	. 12
Aluminium oxide, Al_2O_3	3. 65
Ammonia, NH_3...	1. 02
Sulphuric acid, SO_3	10. 11
Starch..	45. 41
Water of combination and association by difference	24. 80
	100. 00

PROBABLE PERCENTAGE COMBINATION.

Sodium bicarbonate, $NaHCO_3$		18. 98
(Residual sodium oxide, Na_2O)...................................		1. 43
Ammonia alum, anhydrous:		
$Al_2(SO_4)_3$	12. 15	
$(NH_4)_2SO_4$	2. 65	
NH_3	. 34	
		15. 14
Starch..		45. 41
Water of association and of crystallization (alum)...................		19. 04
		100. 00

5528.—*Metropolitan Baking-Powder.*

[Made by Metropolitan Perfume Company, Washington, D. C.]

Available carbonic acid ....per cent..	**8.10**
Cubic inches per ounce of powder at 212° F...............................	97. 3

PERCENTAGE COMPOSITION.

Total carbonic acid, CO_2	9. 45
Sodium oxide, Na_2O..	9. 52
Aluminium oxide, Al_2O_3	3. 73
Ammonia, NH_3...	1. 07
Sulphuric acid, SO_3	10. 71
Starch..	43. 25
Water of combination and association by difference	22. 27
	100. 00

Sodium bicarbonate, $NaHCO_3$	18.04
(Residual sodium oxide, Na_2O)	2.86
Ammonia alum, anhydrous:	
$Al_2(SO_4)_3$ 12.42	
$(NH_4)_2SO_4$ 3.33	
NH_3 .... .21	
	15.96
Starch..	43.25
Water of association and crystallization (alum)..............................	19.89
	100.00

5531.—Cottage Baking-Powder.

[Made by New York Yeast Company, New York.]

Available carbonic acid ....per cent..	**6.62**
Cubic inches per ounce of powder at 212° F.......................................	79.5

PERCENTAGE COMPOSITION.

Total carbonic acid, CO_2	**7.80**
Sodium oxide, Na_2O..	6.77
Aluminium oxide, Al_2O_3	3.92
Ammonia, NH_3	.94
Sulphuric acid, SO_3..	10.63
Starch..	52.29
Water of combination and association by difference	20.10
	100.00

PROBABLE PERCENTAGE COMBINATION.

Sodium bicarbonate, $NaHCO_3$...	14.89
(Residual sodium oxide, Na_2O)	1.28
Ammonia alum, anhydrous:	
$Al_2(SO_4)_3$ 14.05	
$(NH_4)_2SO_4$ 3.47	
NH_3 .... .32	
	17.84
Starch...	52.29
Water of association and crystallization (alum)...............................	13.70
	100.00

MIXED POWDERS.

5514.—Dooley's Baking-Powder.

[Made by Dooley & Bro., New York.]

Available carbonic acid...per cent..	**9.62**
Cubic inches per ounce of powder at 212° F.....................................	115.6

PERCENTAGE COMPOSITION.

Total carbonic acid, CO_2	9.55
Sodium oxide, Na_2O	10.31
Potassium oxide, K_2O	4.51
Ammonia, NH_3	.22
Aluminium oxide, Al_2O_3	3.25
Calcium oxide, CaO	.35
Sulphuric acid, SO_3	7.85
Tartaric acid, $C_4H_4O_5$	(?)
Starch	31.54
Water of combination and association (including tartaric acid) by difference	32.42
	100.00

In this and the following sample no satisfactory estimation of the tartaric acid could be obtained in presence of the alum. Both contain free tartaric acid, and the potash present is probably combined as bitartrate. This sample contains a trace of phosphoric acid. This form of mixed powder should undoubtedly be condemned.

5523.—*Miles' Premium Baking-Powder.*

[Made by Joseph H. Larzelere & Co., Philadelphia, Pa.]

Available carbonic acid	per cent..	**3.56**
Cubic inches per ounce of powder at 212° F		42.8

PERCENTAGE COMPOSITION.

Total carbonic acid, CO_2	3.43
Sodium oxide, Na_2O	10.05
Potassium oxide, K_2O	4.78
Calcium oxide, CaO	.28
Ammonia, NH_3	.73
Aluminium oxide, Al_2O_3	3.59
Tartaric acid, $C_4H_4O_5$	(?)
Sulphuric acid, SO_3	9.65
Starch	18.72
Water of combination and association (including tartaric acid) by difference	48.77
	100.00

ALUM AND PHOSPHATE POWDERS.

5510.—*Henkels' Baking-Powder.*

[Made by Henkels Bros., Paterson, N. J.]

Available carbonic acid	per cent..	**7.71**
Cubic inches per ounce of powder at 212° F		93.0

Total carbonic acid, CO_2	8.34
Sodium oxide, Na_2O	12.25
Aluminium oxide, Al_2O_3	3.57
Calcium oxide, CaO	1.80
Phosphoric acid, P_2O_5	5.60
Sulphuric acid, SO_3	9.79
Ammonia, NH_3	.86
Starch	40.91
Water of combination and association by difference	16.88
	100.00

PROBABLE PERCENTAGE COMBINATION.

Sodium bicarbonate, $NaHCO_3$		15.92
(Residual sodium oxide, Na_2O)		6.38
Ammonia alum, anhydrous:		
$Al_2(SO_4)_3$	11.89	
$(NH_4)_2SO_4$	2.42	
NH_3	.24	
		14.55
Acid phosphate of lime, anhydrous:		
$Ca_3(PO_4)_2$	3.32	
H_3PO_4	5.63	
		8.95
Starch		40.91
Water of association (phosphate) and crystallization (alum)		13.29
		100.00

5511.—*Mason's Yeast-Powder.*

[Made by the Dixon Yeast-Powder Company, 231 Seventh street S. W., Washington, D. C.]

Available carbonic acid ....per cent..	**9.96**
Cubic inches per ounce of powder at 212° F	119.6

PERCENTAGE COMPOSITION.

Total carbonic acid, CO_2	10.66
Sodium oxide, Na_2O	12.53
Aluminium oxide, Al_2O_3	4.27
Calcium oxide, CaO	1.17
Phosphoric acid, P_2O_5	3.58
Sulphuric acid, SO_3	11.02
Ammonia, NH_3	1.04
Starch	43.83
Water of combination and associations by differences.	11.85
	100.00

PROBABLE PERCENTAGE COMBINATION.

Sodium bicarbonate, $NaHCO_3$ 20.34
(Residual sodium oxide, Na_2O) 5.07
Ammonia alum, anhydrous:
    $Al_2(SO_4)_3$ 14.22
    $(NH_4)_2SO_4$ 1.76
    NH_3 .... .59
               —— 16.57
Acid phosphate of lime, anhydrous:
    $Ca_3(PO_4)_2$ 2.16
    H_3PO_4 .... 3.57
               —— 5.73
Starch 43.83
Water of association (phosphate) and crystallization (alum)............... 8.46

                         100.00

5512.—Dixon Yeast-Powder.

[Made by the Dixon Yeast-Powder Company, 231 Seventh street. S. W., Washington, D. C.]

Available carbonic acid ....per cent.. **10.37**
Cubic nches per ounce of powder at 212° F................................ 124.6

PERCENTAGE COMPOSITION.

Total carbonic acid, CO_2 .... 10.68
Sodium oxide, Na_2O 14.04
Calcium oxide, CaO 1.29
Aluminium oxide, Al_2O_3 .... 4.59
Ammonia, NH_3 .... 1.13
Phosphoric acid, P_2O_5 .... 3.38
Sulphuric acid, SO_3 .... 11.57
Starch 42.93
Water of combination and association by difference........................ 10.39

                         100.00

PROBABLE PERCENTAGE COMBINATION.

Sodium bicarbonate, $NaHCO_3$ 20.39
(Residual sodium oxide, Na_2O) 6.52
Ammonia alum anhydrous:
    $Al_2(SO_4)_3$ 15.28
    $(NH_4)_2SO_4$ 1.67
    NH_3 .... .76
               —— 17.71
Acid phosphate of lime, anhydrous:
    $Ca_3(PO_4)_2$ 2.38
    H_3PO_4 .... 3.16
               —— 5.54
Starch 42.93
Water of association (phosphate) and crystallization (alum)............... 6.91

                         100.00

This and the preceding sample are made by the same firm, and are very similar in composition.

5515.—*Patapsco Baking-Powder.*

[Made by Smith, Hanway & Co., Baltimore, Md. Put up in glasses.]

Available carbonic acid ....per cent..	**7.58**
Cubic inches per ounce of powder at 212° F................................	91.1

PERCENTAGE COMPOSITION.

Total carbonic acid, CO_2	9.18
Sodium oxide, Na_2O	9.83
Aluminium oxide, Al_2O_3	4.55
Calcium oxide, CaO	2.77
Phosphoric acid, P_2O_5	1.44
Sulphuric acid, SO_3	13.01
Ammonia, NH_3	.91
Starch	41.24
Water of combination and association by difference.......................	17.07
	100.00

PROBABLE PERCENTAGE COMBINATION.

Sodium bicarbonate, $NaHCO_3$		17.52
(Residual sodium oxide Na_2O)		3.36
Ammonia alum, anhydrous:		
$Al_2(SO_4)_3$	15.15	
$(NH_4)_2SO_4$	3.53	
		18.68
Calcium sulphate, $CaSO_4$		.46
Acid phosphate of lime, anhydrous:		
CaO	2.58	
P_2O_5	1.44	
		4.02
Starch		41.24
Water of association (phosphate) and crystallization (alum) .		14.72
		100.00

This and the three samples following all have the same brand, but were put up in different shape, No. 5515 being contained in glass bottles, No. 5516 in tin cans, and No. 5517 being sold in bulk. No. 5519 is another brand made by the same firm. Of these, Nos. 5515 and 5517 give very similar results on analysis, while in Nos. 5516 and 5519 the numbers agree closely with one another, though quite different from the other two samples. In Nos. 5515 and 5517 the acid phosphate of lime is given simply as the sum of the per cents. of CaO and P_2O_5 found. There is probably considerable calcium sulphate in these samples, but if the sulphuric acid is combined with all the alum there is not enough left for combination with the lime.

5516.—*Patapsco Baking-Powder.*

[Made by Smith, Hanway & Co., Baltimore, Md. In tin cans.]

Available carbonic acid ....per cent..	**6.70**
Cubic inches per ounce of powder at 212° F........	55.16

PERCENTAGE COMPOSITION.

Total carbonic acid, CO_2........	7.62
Sodium oxide, Na_2O	9.21
Aluminium oxide, Al_2O_3........	3.61
Calcium oxide, CaO	2.66
Ammonia, NH_3........	.88
Phosphoric acid, P_2O_5........	3.23
Sulphuric acid, SO_3........	12.26
Starch........	36.39
Water of combination and association by difference........	24.09
	100.00

PROBABLE PERCENTAGE COMBINATION.

Sodium bicarbonate, $NaHCO_3$		14.54
(Residual sodium oxide, Na_2O)........		3.84
Ammonia alum, anhydrous:		
$Al_2(SO_4)_3$........	12.02	
$(NH_4)_2SO_4$	3.41	
		15.43
Calcium sulphate, $CaSO_4$........		3.02
Acid phosphate of lime, anhydrous:		
$Ca_3(PO_4)_2$........	2.62	
H_3PO_4	2.87	
		5.49
Starch........		36.39
Water of association (phosphate) and crystallization (alum)		21.29
		100.00

5517.—*Patapsco Baking-Powder.*

[Made by Smith, Hanway & Co., Baltimore, Md. Sold in bulk.]

Available carbonic acid ....per cent..	**8.42**
Cubic inches per ounce of powder at 212° F........	101.1

PERCENTAGE COMPOSITION.

Total carbonic acid, CO_2........	9.55
Sodium oxide, Na_2O........	8.33
Aluminium oxide, Al_2O_3........	4.72
Calcium oxide, CaO........	2.66
Ammonia, NH_3........	.86
Phosphoric acid, P_2O_5........	1.78
Sulphuric acid, SO_3........	13.18
Starch	43.92
Water of combination and association by difference........	15.00
	100.00

Sodium bicarbonate, NaHCO₃		18.21
(Residual sodium oxide, Na₂O)		1.60
Ammonia alum, anhydrous:		
Al₂(SO₄)₃	15.72	
(NH₄)₂SO₄	3.34	
		19.06
Acid phosphate of lime, anhydrous:		
CaO	2.66	
P₂O₅	1.78	
		4.44
Starch		43.92
Water of association (phosphate) and crystallization (alum)		12.77
		100.00

5519.—*Silver Spoon Baking-Powder.*

[Made by Smith, Hanway & Co., Baltimore, Md.]

Available carbonic acid .·.............per cent..	**7.33**
Cubic inches per ounce of powder at 212° F.	88.0

PERCENTAGE COMPOSITION.

Total carbonic acid, CO₂	8.26
Sodium oxide, Na₂O	7.26
Potassium oxide, K₂O	.76
Calcium oxide, CaO	2.29
Aluminium oxide, Al₂O₃	3.58
Ammonia, NH₃	.91
Phosphoric acid, P₂O₅	3.61
Sulphuric acid, SO₃	8.78
Starch	41.26
Water of combination and association by difference	23.29
	100.00

PROBABLE PERCENTAGE COMBINATION.

Sodium bicarbonate, NaHCO₃		16.13
(Residual sodium oxide, Na₂O)		1.44
Ammonia alum, anhydrous:		
Al₂(SO₄)	11.92	
(NH₄)₂SO₄	.72	
NH₃	.78	
		13.42
Acid phosphate of lime, anhydrous:		
Ca₃(PO₄)₂	4.22	
H₃PO₄	2.32	
		6.54
Starch		41.26
Water of association (phosphate) and crystallization (alum)		21.21
		100.00

5520.— Windsor Baking-Powder.

[Made by Edwin J. Gillies & Co., 245 and 247 Washington street, New York.]

Available carbonic acid ....per cent.. **9.36**
Cubic inches per ounce of powder at 212° F.. 112.4

PERCENTAGE COMPOSITION.

Total carbonic acid, CO_2	9.86
Sodium oxide, Na_2O...	11.92
Calcium oxide, CaO...	1.76
Ammonia, NH_3...	.99
Aluminium oxide Al_2O_3..	3.60
Phosphoric acid, P_2O_5	5.01
Sulphuric acid, SO_3	10.51
Starch..	41.26
Water of combination and association by difference...................	15.09
	100.00

PROBABLE PERCENTAGE COMBINATION.

Sodium bicarbonate. $NaHCO_3$		18.82
(Residual sodium oxide, Na_2O)		4.98
Ammonia alum, anhydrous:		
$Al_2(SO_4)_3$	11.99	
$(NH_4)_2SO_4$..	3.50	
		15.49
Acid phosphate of lime, anhydrous:		
$Ca_2(PO_4)_2$..	3.25	
H_3PO_4	4.86	
		8.11
Starch..		41.26
Water of association (phosphate) and crystallization (alum)		11.34
		100.00

5521.— Davis' O. K. Baking-Powder.

[Made by R. B. Davis, 112 Murray street, New York.]

Available carbonic acid ....per cent.. **8.10**
Cubic inches per ounce of powder at 212° F.................................... 97.3

PERCENTAGE COMPOSITION.

Total carbonic acid, CO_2..	9.02
Sodium oxide, Na_2O..	11.20
Calcium oxide, CaO..	3.47
Ammonia, NH_3..	1.04
Aluminium oxide, Al_2O_3...	4.67
Phosphoric acid, P_2O_5..	8.95
Sulphuric acid, SO_3	11.54
Starch...	32.85
Water of combination and association by difference....................	17.26
	100.00

Sodium bicarbonate, $NaHCO_3$ 17.22
(Residual sodium oxide, Na_2O) 4.85
Ammonia alum, anhydrous:
    $Al_2(SO_4)_3$ 15.55
    $(NH_4)_2SO_4$ 1.09
    NH_3 .... .76
                          ——— 17.40
Acid phosphate of lime, anhydrous:
    $Ca_3(PO_4)_2$ 6.40
    H_3PO_4.. 8.31
                          ——— 14.71
Starch... 32.85
Water of association (phosphate) and crystallization (alum).............. 12.97

                             100.00

5524.—*Brunswick Yeast-Powder.*

[Manufactured in New York for M. &. P. Metzger, 417 Seventh street, Washington, D. C.]

Available carbonic acid..per cent.. **9.81**
Cubic inches per ounce of powder at 212° F................................... 117.8

PERCENTAGE COMPOSITION.

Total carbonic acid, CO_2 .... 11.49
Sodium oxide, Na_2O... 10.87
Calcium oxide, CaO.. 2.22
Aluminium oxide, Al_2O_3.. 3.35
Ammonia, NH_3... 1.59
Phosphoric acid, P_2O_5... 5.11
Sulphuric acid, SO_3.. 10.14
Starch.. 34.97
Water of combination and association by difference........................... 20.26

                             100.00

PROBABLE PERCENTAGE COMBINATION.

Sodium bicarbonate, $NaHCO_3$... 21.93
(Residual sodium oxide, Na_2O)... 2.78

Ammonia alum, anhydrous:
    $Al_2(SO_4)_3$ 11.15
    $(NH_4)_2SO_4$... 3.86
    NH_3 .... .60
                          ——— 15.61
Acid phosphate of lime, anhydrous:
    $Ca_3(PO_4)_2$.. 3.56
    H_3PO_4 .... 5.63
                          ——— 9.19
Starch.. 34.97
Water of association (phosphate) and crystallization (alum)............... 15.52

                             100.00

5525.—*The Atlantic and Pacific Baking-Powder.*

[Made by The Atlantic & Pacific Tea Company, New York.]

Available carbonic acid...per cent..	**7. 91**
Cubic inches per ounce of powder at 212° F...............................	95. 0

PERCENTAGE COMPOSITION.

Total carbonic acid, CO_2..	9. 45
Sodium oxide, Na_2O..	12. 15
Calcium oxide, CaO..	1. 93
Aluminium oxide, Al_2O_3	3. 25
Ammonia, NH_3..	.72
Phosphoric acid, P_2O_5...	5. 71
Sulphuric acid, SO_3..	8. 93
Starch..	37. 66
Water of combination and association by difference......................	20. 20
	100. 00

PROBABLE PERCENTAGE COMBINATION.

Sodium bicarbonate, $NaHCO_3$		18. 04
(Residual sodium oxide, Na_2O)		5. 15
Ammonia alum, anhydrous :		
$Al_2(SO_4)_3$...	10. 82	
$(NH_4)_2 SO_4$...	2. 41	
		13. 23
Acid phosphate of lime, anhydrous:		
$Ca_3(PO_4)_2$	3. 56	
H_3PO_4	5. 63	
		9. 19
Starch		37. 66
Water of association (phosphate) and crystallization (alum)		16. 73
		100. 00

5530.—*Silver King Baking-Powder.*

[Made by Shaw & Thomas, New York.]

Available carbonic acid..per cent..	**4. 99**
Cubic inches per ounce of powder at 212° F...............................	59. 9

PERCENTAGE COMPOSITION.

Total carbonic acid, CO_2	6. 12
Sodium oxide, Na_2O...	10. 32
Calcium oxide, CaO...	2. 79
Ammonia, NH_3..	1. 04
Aluminium oxide, Al_2O_3...	3. 75
Phosphoric acid, P_2O_5	5. 89
Sulphuric acid, SO_3	10. 57
Starch...	42. 66
Water of combination and association by difference.....................	16. 86
	100. 00

Sodium bicarbonate, $NaHCO_3$ 11.68
(Residual sodium oxide, Na_2O) 6.01
Ammonia alum, anhydrous:
    $Al_2(SO_4)_3$ 12.49
    $(NH_4)_2SO_4$ 3.02
    NH_3 .... .26
                    15.77
Acid phosphate of lime, anhydrous:
    $Ca_3(PO_4)_2$ 5.15
    H_3PO_4 .... 4.87
                    10.02
Starch 42.66
Water of association (phosphate) and crystallization (alum) 13.86

                    100.00

5532.—*Eureka Baking-Powder.*

[Made by G. S. Feeny, Wheeling, W. Va.]

Available carbonic acid ....per cent.. **7.62**
Cubic inches per ounce of powder at 212° F 91.5

PERCENTAGE COMPOSITION.

Total carbonic acid, CO_2 .... 9.57
Sodium oxide, Na_2O 11.36
Calcium oxide, CaO 1.93
Aluminium oxide, Al_2O_3 .... 3.14
Ammonia, NH_3 .... .85
Phosphoric acid, P_2O_5 .... 2.66
Sulphuric acid, SO_3 .... 11.30
Starch 44.32
Water of combination and association by difference 14.87

                    100.00

PROBABLE PERCENTAGE COMBINATION.

Sodium bicarbonate, $NaHCO_3$ 18.27
(Residual sodium oxide, Na_2O) 4.62
Ammonia alum, anhydrous:
    $Al_2(SO_4)_3$ 10.45
    $(NH_4)_2SO_4$ 3.31
                    13.76
Calcium sulphate, $CaSO_4$ 3.38
Acid phosphate of lime, anhydrous:
    $Ca_3(PO_4)_2$ .... .99
    H_3PO_4 .... 2.92
                    3.91
Starch 44.32
Water of association (phosphate) and crystallization (alum) 11.74

                    100.00

5533.—*Silver Star Baking-Powder.*

[Made by E. Canby, Dayton, Ohio.]

Available carbonic acid...per cent.. **7.61**
Cubic inches per ounce of powder at 212° F.................................... 91. 4

PERCENTAGE COMPOSITION.

Total carbonic acid, CO_2.....................................	9. 89
Sodium oxide, Na_2O...	12. 69
Potassium oxide, K_2O	.74
Calcium oxide, CaO......................................	2. 16
Aluminum oxide, Al_2O_3.....................................	3. 38
Ammonia, NH_3..	.77
Phosphoric acid, P_2O_5.....................................	5. 30
Sulphuric acid, SO_3......................................	10. 66
Starch..	37. 57
Water of combination and association by difference	16. 84
	100. 00

PROBABLE PERCENTAGE COMBINATION.

Sodium bicarbonate, $NaHCO_3$....		18. 88
(Residual sodium oxide, Na_2O)......................................		5. 73
Ammonia alum, anhydrous :		
$Al_2(SO_4)_3$...	11. 25	
$(NH_4)_2SO_4$	2. 98	
		14. 23
Calcium sulphate, $CaSO_4$.......................................		1. 66
Acid phosphate of lime, anhydrous:		
$Ca_3(PO_4)_2$	2. 73	
H_3PO_4	5. 59	
		8. 32
Starch...		37. 57
Water of association (phosphate) and crystallization (alum)................		13. 61
		100. 00

5534.—*Purity Baking-Powder.*

[Made by Smith, Hanway & Co., Baltimore, Md.]

Available carbonic acid ....per cent.. **7.13**
Cubic inches per ounce of powder at 212° F....................................... 85. 6

PERCENTAGE COMPOSITION.

Total carbonic acid, CO_2.....................................	9. 32
Sodium oxide, Na_2O...	12. 99
Calcium oxide, CaO	4. 55
Aluminum oxide, Al_2O_3	3. 34
Ammonia, NH_3..	.93
Phosphoric acid, P_2O_5	3. 53
Sulphuric acid, SO_3......................................	13. 23
Starch..	40. 19
Water of combination and association by difference	12. 62
	100. 00

PROBABLE PERCENTAGE COMBINATION.

Sodium bicarbonate, $NaHCO_3$		18.75
(Residual sodium oxide, Na_2O)		5.37
Ammonia alum, anhydrous:		
$Al_2(SO_4)_3$	11.12	
$(NH_4)_2SO_4$	3.59	
		14.71
Calcium sulphate, $CaSO_4$		5.58
Acid phosphate of lime, anhydrous:		
$Ca_3(PO_4)_2$	4.19	
H_3PO_4	2.51	
		6.70
Starch		40.19
Water of association (phosphate) and crystallization (alum)		8.70
		100.00

These analyses agree pretty closely, in a general way, with those made by Professors Weber and Cornwall, so far as their determinations go. Professor Cornwall's figures for carbonic acid are uniformly higher than mine, his method giving rather the total than the available per cent. of gas. It is evident that the determination of available carbonic acid made upon samples obtained in retail stores would vary more or less, according to the time that had elapsed since the sample was first put up. This is well shown by Professor Cornwall's determinations (page 586), made upon samples of the same brand of powder purchased at different times. It would manifestly be unjust, therefore, to decide arbitrarily that the relative values of the different brands were in the exact rank indicated by the results given in this determination. The best results in all the different investigations, however, are given by the tartrate and phosphate powders, the alum, and the alum and phosphate powders giving almost uniformly low percentages of carbonic acid. (There are several exceptions, however, notably Professor Cornwall's No. 36, "One Spoon" giving 16.77 per cent.[1] This shows the possibilities of an alum powder as regards its carbonic-acid strength.)

Professor Cornwall's average for twenty samples of alum and phosphate powders (no straight alum powders included) is 8.97 per cent.; for eight samples of tartrate powders, 11.60. Professor Weber's average for nineteen samples of alum powders is 7.58 per cent.; for eight samples of tartrate powders, 11.20 per cent. My average for twenty samples of both alum, and alum and phosphate powders is 8 per cent.; for eight samples of tartrate powders, 10.10 per cent. The only straight phosphate powders sold seem to be the various preparations made by the Rumford Chemical Works, and the "Wheat" powder; at least these are all obtained by any of the investigators. The carbonic acid strength of the former is uniformly good, slightly higher than the tartrate pow-

---

[1] Weber obtained only 5.75 per cent. from a powder with this brand in his investigation. Page —.

ders; the latter is a peculiar preparation, made up without any filling whatever, and gives a very low percentage of carbonic acid, except in one of Professor Cornwall's samples, which seems to have been obtained quite fresh.

FILLING.

It is evident that of several powders made up of the same materials, the one which contains the smallest proportion of inert matter or filling, other things being equal, will have the best carbonic-acid efficiency or "strength." On the other hand, if the amount used is too small for the proper preservation of the sample, it will deteriorate rapidly, and perhaps will show less strength after keeping a short time than other powders with a somewhat larger amount of filling. It becomes a question, therefore, as to the minimum limit of the amount of filling that is consistent with good keeping qualities. Professor Prescott[1] says on this point:

From 13 to 18 per cent. of starch is not too much for the permanence of a cream of tartar baking-powder, but filling beyond 20 per cent. must be held an unquestionable dilution.

In my samples, the average per cent. of starch in the bitartrate powders was 14.04; the highest was 24.57 per cent., and the lowest 5.32 per cent. The latter sample evidently did not contain enough, for it had a much lower carbonic-acid strength than most of those that had more filling. The bitartrate powder containing the maximum of filling, No. 5527, contained also the lowest per cent. of available carbonic acid. The powders made up with free tartaric acid contained much more filling, this being doubtless necessitated by the more hygroscopic character of the free acid. They contain, respectively, 40.05 and 45.63 per cent. of starch, and 9.53 and 4.98 per cent. of available carbonic-acid. Of the phosphate powders No. 5508 contains rather a large amount of filling, 26.41 per cent., while No. 5506 contains none at all, evidently to its detriment, as previously noted. Even the acid part of No. 5509 contains 20.81 per cent. of starch, although it is kept separate from the alkali. It is in the alum, and the alum and phosphate powders, however, that the highest percentages of filling are found. The average of all is 40.76 per cent. of starch, the maximum 52.29 per cent., the minimum, 31.54 per cent. Here we find the cause for the low per cents. of available carbonic acid in these powders, which should, theoretically, afford a higher carbonic-acid strength than any of the other classes. Whether a large amount of filling is more necessary where alum is used to prevent deterioration, whether it is added simply as a diluent, so that the amount of alum taken into the starch will be less apt to produce an injurious effect, or whether it is added to cheapen the powder, I can not say. The first hypothesis seems the most probable,

---

[1] Organic Analyses, 500.

especially if the alum is used with but a small proportion of its water of crystallization driven off. If the second is true, the object is not obtained, of course, for the more filling used the greater the quantity of powder required to produce the same aerating effect, and as for the third, alum and soda are about as cheap as starch.

It must be remembered that the percentages of starch given in the tables represent *anhydrous* starch.

"DOMESTIC BAKING-POWDERS."

It may be asked, can not the consumer make up his own baking-powders? The difficulties in the way of doing this may be enumerated as follows:

(1) The chemicals in the market, as purchased by the consumer, may not be pure, or of full strength, so that when combined in proper proportions they do not give good results.

(2) The proper proportions to use, and the necessity of thorough mixing to secure good results, would not be well understood by any one who was not a chemist.

(3) In order to prevent the action of the ingredients upon one another, and to preserve the strength of the powder unimpaired as long as possible, the manufacturer *dries* all his chemicals before mixing them, so as to drive off most of the adhering moisture. Baking-soda can not be dried much, as it loses its carbonic acid, and consequently its efficiency, at very low temperatures. The starch, however, containing as it does from 10 to 18 per cent. of moisture, can be thoroughly dried at 100° to 105° C., and its efficiency as a filling material greatly increased. The cream of tartar can also be thoroughly dried. This operation of drying chemicals at a temperature below that at which decomposition would occur seems rather too elaborate an operation for the kitchen.

These difficulties are more apparent than real, however. In answer to the first, it may be said that the bitartrate is the only chemical which is likely to be adulterated, and as there is no difficulty nowadays in obtaining a pure article in the wholesale market, it only requires the proper enforcement of adulteration laws to oblige the retailer to furnish a good article. The second objection may be met by furnishing the public simple formulæ for compounding such powders, and the third, which is doubtless the most serious, I believe can be overcome by using a larger proportion of filling, without drying the chemicals.

In the present days of cooking-schools, when so much interest is taken in the preparation of food, and in all branches of the culinary art, it may not be amiss to devote a little space to the discussion of this subject, although it is not, perhaps, strictly within the scope of the present investigation.

With a view of determining the possibility of making up baking-powders from a simple formula that could be used in the household, and also to see what strength of powder could be obtained by lessening the quantity of filling used, I compounded a number of powders from commercial cream of tartar and soda, using different proportions of starch, and determined the per cent. of carbonic acid, both total and available, in each. The chemicals used were dried before mixing, and the latter operation very thoroughly performed.

Formula No. 1, containing 20 per cent. starch filling.

Cream of tartar...ounces..   3
Baking-soda..do....   4
Corn starch..do....   3

Total carbonic acid ....per cent..   13.39
Available carbonic acid..do....   11.96

Formula No. 2, containing about 15 per cent. starch filling.

Cream of tartar...ounces..   6
Baking-soda..do....   4
Corn starch..do....   2

Total carbonic acid ....per cent..   14.60
Available carbonic acid..do....   12.29

Formula No. 3, containing 10 per cent. starch filling.

Cream of tartar...ounces..   6
Baking-soda..do....   3
Corn starch..do....   1

Total carbonic acid ....per cent..   15.10
Available carbonic acid..do....   13.70

From the above it will be seen that most excellent results were obtained with these powders, made up by simple formulæ. The powder containing the least percentage of starch, Formula No. 3, gave 13.70 per cent. of available carbonic acid, nearly 1 per cent. more than the highest result obtained in any of the commercial samples. To be sure these powders were freshly made, and would doubtless deteriorate on keeping, those with the lowest amount of starch perhaps more rapidly than the others, as most of the commercial samples containing less than 10 per cent. of starch show low percentages of available carbonic acid, No. 5505 being an exception. But these prepared samples establish very completely the point I desired to make, that baking-powders can be readily made up by simple formulæ that will compare favorably with the best samples obtainable in the market.

These samples, however, were all made with *well-dried* ingredients, as they would be by a manufacturer. The next question is, whether a powder could be made which would keep without serious deterioration.

without drying the chemicals. To this end I used a larger proportion of starch according to the following formula:

Formula No. 4, made without drying the ingredients, containing 25 per cent. starch filling.

Cream of tartar .... ounces.. 8
Baking-soda... do.... 4
Corn starch .... do.... 4

Total carbonic acid... per cent.. 12. 63
Available carbonic acid... do..... 10. 91

This gives a fairly good amount of available gas, considerably higher than the average of the commercial samples. Estimations of the available carbonic acid in the same sample after it had stood over two months in the laboratory showed absolutely no loss in strength. I had it tried in a practical way by several persons in the Department who used it in their kitchens, and reported excellent results, finding it fully as efficient in all respects as the powder they were accustomed to buy. The consumer can pay full retail price for the ingredients and still make it up for about half the price at which a good powder is sold, and if he makes sure of the quality of his cream of tartar he will have an article of which the purity is assured, and which has not lost in strength by being kept in stock an indefinite length of time by the retailer. I can see no reason why all housekeepers should not make their own baking-powder.

REGULATION OF THE SALE OF BAKING-POWDERS.

The best plan for the regulation by law of the sale of baking-powders in the present condition of our knowledge of their effect upon the system would seem to be to require the manufacturer to use a label giving approximately the composition, or analysis, of the powder sold. This is recommended by Professor Cornwall, and it appears to offer the best solution of the whole problem. The testimony that has been adduced is hardly sufficient to justify the prohibition of the sale of the cheaper kinds of powders as being injurious to health, but if they were required to be sold with a label giving their true composition it would soon lead to investigations upon this point. This is in harmony, also, with modern ideas in regard to legal regulation of the sale of food-stuffs, the tendency nowadays being to allow the sale of cheap substitutes for any article of food so long as they are not actually injurious to health, but to make all possible provision to insure that the purchaser should know exactly what he is getting, and that the substitute shall not be palmed off on him as the genuine article. In the case of baking-powders it is manifestly unjust to the public to allow the sale of a first-class tartrate powder and an alum powder as the same article, and it is equally unjust to the manufacturer of the higher-priced article. The nature of the substance is such that the purchaser has no means of ascertaining by any simple or easy means the character of the article he buys, to say nothing of its relative quality. Such a regulation should meet with the approbation of all concerned in the manufacture of baking-powders. The manufacturers of high-grade powders, such as tartrate or phosphate powders, would certainly not object to it, and it would ultimately be to the advantage also of the cheaper sorts, such as alum powders, provided they could succeed in proving that such powders produced little or no injury to the health of the consumer.

Ample analogy and precedent for such regulation are furnished by the laws for the sale of fertilizers which are in operation in most of the States. Although these substances are used for widely different purposes, the conditions that require the legal supervision of their sale are quite similar in many respects. A substance sold as a fertilizer must have its composition, in so far as is necessary for its valuation for such a purpose, plainly stated on the bag in which it is sold, because the purchaser has no means of ascertaining this value by any ordinary or simple test. Otherwise the manufacturer could easily impose upon him by selling him a powdered substance which resembled a fertilizer in general appearance, but contained no constituent of any value whatever for fertilizing purposes. The purchaser of a baking-powder receives a white powder which may contain various substances more or less valuable for the desired purpose, or of no value whatever, or perhaps even injurious to the health.

The housewife surely deserves protection against swindling as much as the farmer, and she has no better means for ascertaining the strength and quality of the baking-powder she buys than the latter has for learning the strength of his fertilizer. The verity and accuracy of the analysis stated on the label should be insured, as in the case of the fertilizer, by its being performed by sworn analysts. If such a regulation were enforced, people would soon inform themselves of the respective merits of different varieties, and the further requirement of a certain standard of strength, as suggested by Professor Cornwall, would probably be unnecessary, as they would learn to interpret the analysis, and a powder made up with 50 per cent. of starch, for instance, would have to be sold cheaper than one made with 10 per cent., or not sold at all.

INDEX.

· A.

B.

○

www.ingramcontent.com/pod-product-compliance
Lightning Source LLC
Chambersburg PA
CBHW021532270326
41930CB00008B/1216